STUDENT SOLUTIONS MANUAL

TO ACCOMPANY

FUNCTIONS MODELING CHANGE
A Preparation for Calculus

Eric Connally
Wellesley College

Deborah Hughes-Hallett
University of Arizona

Andrew M. Gleason
Harvard University

et al.

JOHN WILE

NEW YORK • CHICHESTER • WEINHEIM • BRISBANE • SING

COVER PHOTO © Peter Mathis/Tony Stone Images, New York

This material is based upon work supported by the National Science Foundation under Grant No. DUE-9352905. All royalties from the sale of this book will go toward the furtherance of the project. Opinions expressed are those of the authors and not necessarily those of the Foundation.

To order books or for customer service call 1-800-CALL-WILEY (225-5945).

ISBN 0-471-29396-2

Printed in the United States of America

10 9 8 7 6 5 4 3

Printed and bound by Bradford & Bigelow, Inc.

CONTENTS

CHAPTER ONE

Solutions for Section 1.1

1. (a) 69°F
   (b) July 17th and 20th
   (c) Yes. For each date, there is exactly one low temperature.
   (d) No, it is not true that for each low temperature, there is exactly one date: for example, 73° corresponds to both the 17th and 20th.

5. These data are plotted in Figure 1.1. The independent variable is A and the dependent variable is n.

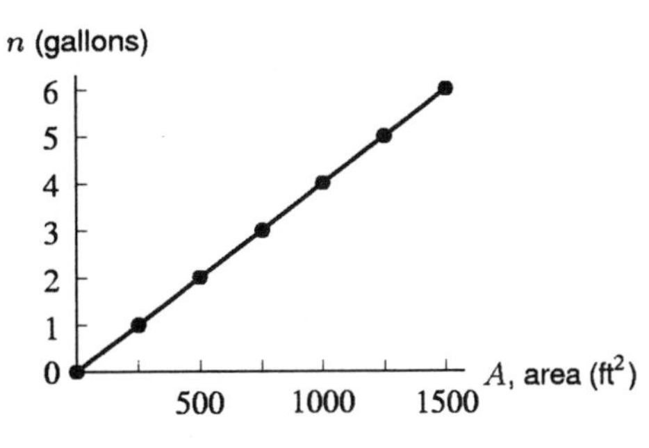

Figure 1.1

9. (a) The graphs in (I), (III), (IV), (V), (VII), and (VIII) are functions. The graphs in (II), (VI), and (IX) do not pass the vertical line test and so they cannot be the graphs of functions.
   (b) (i) The graph of SAT Math score versus SAT Verbal score for a number of students will be a graph of a number of points. Graphs (V) and (VI) are of this type.
   (ii) The graph of hours of daylight per day must be an oscillating function (since the number of hours of daylight fluctuates up and down throughout the year). Graph (VIII) represents this.
   (c) If the train fare remains constant throughout the day, graph (III) describes the fare. If there are specific times of the day (rush hours, for example) when the train company raises its prices, then graph (IV) represents the train fare as a function of time of day.

13. Appropriate axes are shown in Figure 1.2.

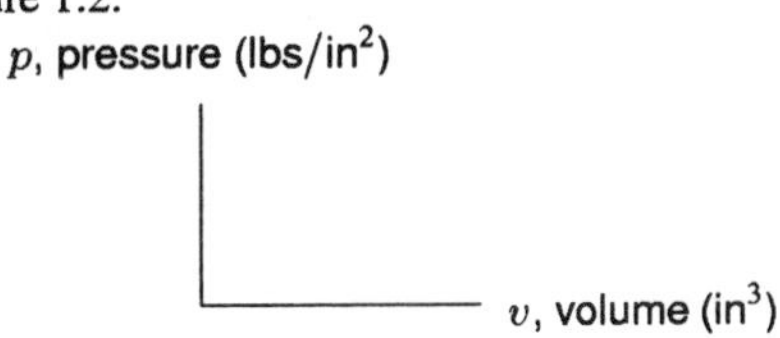

Figure 1.2

17. A possible graph is shown in Figure 1.3.

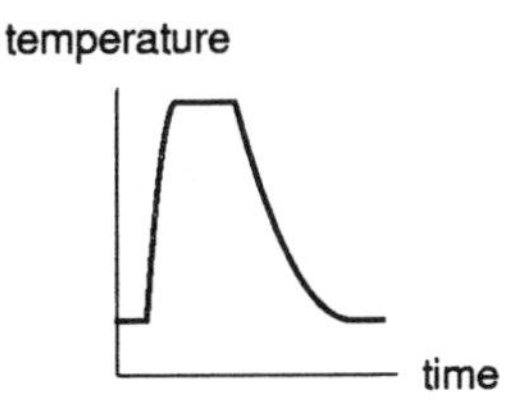

Figure 1.3

21. (a) No, because the same value of x is associated with more than one value of y.
    (b) Yes, because each value of y is associated with exactly one value of x (in this case, $x = 5$).

25. (a) Yes. If the person walks due west and then due north, the distance from home is represented by the hypotenuse of the right triangle that is formed (see Figure 1.4).

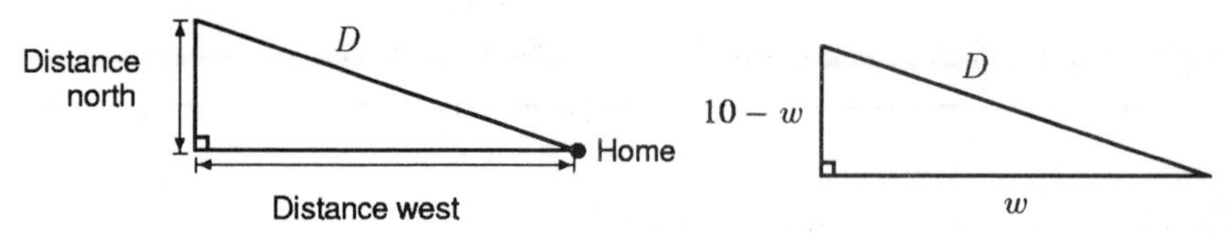

Figure 1.4

If the distance west is w miles and the total distance walked is 10 miles, then the distance north is $10 - w$ miles.

We can use the Pythagorean Theorem to find that

$$D = \sqrt{w^2 + (10 - w)^2}.$$

So, for each value of w, there is a unique value of D given by this formula. Thus, the definition of a function is satisfied.

(b) No. Suppose she walks 10 miles, that is, $x = 10$. She might walk 1 mile west and 9 miles north, or 2 miles west and 8 miles north, or 3 miles west and 7 miles north, and so on. The right triangles in Fig 1.5 show three different routes she could take and still walk 10 miles.

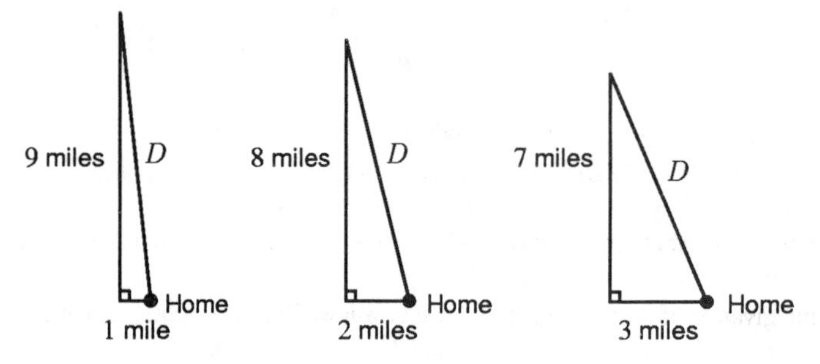

Figure 1.5

Each situation gives a different distance from home. The Pythagorean Theorem shows that the distances from home for these three examples are

$$D = \sqrt{1^2 + 9^2} = 9.06,$$
$$D = \sqrt{2^2 + 8^2} = 8.25,$$
$$D = \sqrt{3^2 + 7^2} = 7.62.$$

Thus, the distance from home cannot be determined from the distance walked.

Solutions for Section 1.2

1. Substituting into the general formula $y = kx$, we have $6 = k(4)$ or $k = \frac{3}{2}$. So the formula for y is

$$y = \frac{3}{2}x.$$

When $y = 8$, we get $8 = \frac{3}{2}x$, so $x = \frac{2}{3} \cdot 8 = \frac{16}{3} = 5.33$.

5. This is a case of direct proportionality.

$$y = -x = (-1)x^{(1)}.$$

Thus $k = -1$ and $p = 1$.

9. This is a case of direct proportionality.

$$y = \frac{(0.34)}{2}x = 0.17x = (0.17)x^{(1)}.$$

Thus $k = 0.17$ and $p = 1$.

13. (a) Since the cost of the fabric, $C(x)$, is directly proportional to the amount purchased, x, we know that the formula will be of the form
$$C(x) = kx.$$
(b) Since 3 yards cost \$28.50, we know that $C(3) = \$28.50$. Thus, we have
$$28.50 = 3k$$
$$k = 9.5$$
Our formula for the cost of x yards of fabric is
$$C(x) = 9.5x.$$
(c) Notice that the graph in Figure 1.6 goes through the origin.

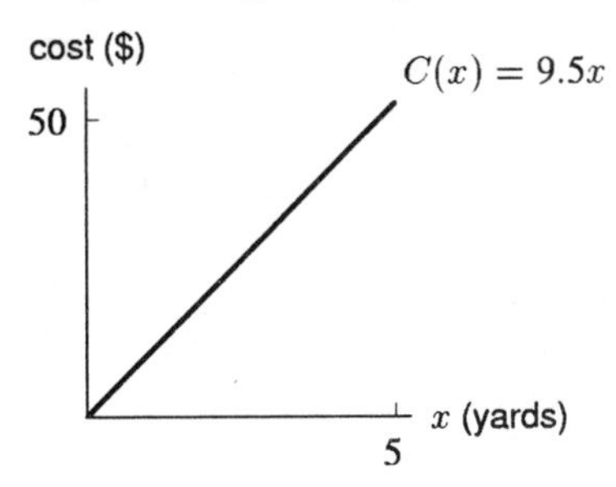

Figure 1.6

(d) To find the cost of 5.5 yards of fabric, we evaluate $C(x)$ for $x = 5.5$:
$$C(5.5) = 9.5(5.5) = \$52.25.$$

17. The number of hours is inversely proportional to the speed, because as the speed, v, increases, the number of hours, h, decreases.

Substituting the given values into the general formula $h = k/v$, we get $3.5 = k/55$, so $k = 192.5$ and the formula is
$$h = \frac{192.5}{v}.$$
When $h = 3$, we get $3 = 192.5/v$, or $v = 64.167$. So getting to Albany in 3 hours would require the speed of 64.167 mph.

21. This function represents proportionality to a power.
$$y = (2x)^5 = (2^5)x^5 = (32)x^{(5)}.$$
Thus, $k = 32$ and $p = 5$.

25. This function represents proportionality to a power.
$$y = \frac{-5}{t^{-4}} = (-5)t^{(4)}.$$
Thus $k = -5$ and $p = 4$.

29. This function does not represent proportionality to a power because we cannot get it into the form $y = kx^p$. Instead, we have a function of the form $y = kp^x$, with constant p.

33. (a) For both circle and sphere, $d = 2r$, so $r = d/2$ and
$$\text{Area of circle} = \pi r^2 = \pi\left(\frac{d}{2}\right)^2 = \pi\frac{d^2}{4}$$
$$\text{Volume of sphere} = \frac{4}{3}\pi r^3 = \frac{4}{3}\pi\left(\frac{d}{2}\right)^3 = \frac{4\pi}{3}\cdot\frac{d^3}{8} = \frac{\pi d^3}{6}.$$
(b) Yes, A is proportional to d^2, with $k = \pi/4$.
Yes, V is proportional to d^3, with $k = \pi/6$.

37. (a) We expect $C = kd^2$, so we calculate this ratio for each of the data points in the table.
$$\frac{C}{d^2} = \frac{3696}{10^2} = 36.96, \quad \frac{14784}{20^2} = 36.96, \quad \frac{33264}{30^2} = 36.96, \quad \frac{59136}{40^2} = 36.96.$$
Since these ratios are constant (all are 36.96), cost appears to be proportional to diameter.

(b) The constant is $k = 36.96$, so
$$C = f(d) = 36.96d^2.$$

(c) The area of a hemisphere (half sphere) is
$$\text{Area} = \frac{1}{2} \cdot 4\pi r^2 = 2\pi r^2.$$
Since the diameter is twice the radius, $d = 2r$, so $r = d/2$. Substituting gives
$$\text{Area} = 2\pi \left(\frac{d}{2}\right)^2 = 2\pi \frac{d^2}{4} = \frac{\pi d^2}{2}.$$
Since 1 ounce of gold covers 17 m^2, for an area of $\pi d^2/2$
$$\text{Number of ounces of gold needed} = \frac{\pi d^2/2}{17} = \frac{\pi d^2}{34}.$$
One ounce of gold costs \$400, so
$$\text{Cost} = 400 \cdot \frac{\pi d^2}{34} = \frac{400\pi d^2}{34}.$$
A calculator gives
$$\frac{400\pi}{34} = 36.96,$$
so
$$C = \frac{400\pi}{34} d^2 = 36.96d^2.$$

Solutions for Section 1.3

1. (a) Let $s = C(t)$ be the sales (in millions) of CDs in year t. Then
$$\begin{aligned}\text{Average rate of change of } s \text{ from } t = 1982 \text{ to } t = 1984 &= \frac{\Delta s}{\Delta t} = \frac{C(1984) - C(1982)}{1984 - 1982} \\ &= \frac{5.8 - 0}{2} \\ &= 2.9 \text{ million discs/year.}\end{aligned}$$
Let $q = L(t)$ be the sales (in millions) of LPs in year t. Then
$$\begin{aligned}\text{Average rate of change of } q \text{ from } t = 1982 \text{ to } t = 1984 &= \frac{\Delta q}{\Delta t} = \frac{L(1984) - L(1982)}{1984 - 1982} \\ &= \frac{205 - 244}{2} \\ &= -19.5 \text{ million records/year.}\end{aligned}$$

(b) By the same argument
$$\begin{aligned}\text{Average rate of change of } s \text{ from } t = 1986 \text{ to } t = 1988 &= \frac{\Delta s}{\Delta t} = \frac{C(1988) - C(1986)}{1988 - 1986} \\ &= \frac{150 - 53}{2} \\ &= 48.5 \text{ million discs/year.}\end{aligned}$$
$$\begin{aligned}\text{Average rate of change of } q \text{ from } t = 1986 \text{ to } t = 1988 &= \frac{\Delta q}{\Delta t} = \frac{L(1988) - L(1986)}{1988 - 1986} \\ &= \frac{72 - 125}{2} \\ &= -26.5 \text{ million records/year.}\end{aligned}$$

(c) The fact that $\Delta s/\Delta t = 2.9$ tells us that CD sales increased at an average rate of 2.9 million discs/year between 1982 and 1984. The fact that $\Delta s/\Delta t = 48.5$ tells us that CD sales increased at an average rate of 48.5 million discs/year between 1986 and 1988.

The fact that $\Delta q/\Delta t = -19.5$ means that LP sales decreased at an average rate of 19.5 million records/year between 1982 and 1984. The fact that the average rate of change is negative tells us that annual sales are decreasing.

The fact that $\Delta q/\Delta t = -26.5$ means that LP sales decreased at an average rate of 26.5 million records/year between 1986 and 1988.

5. No, $f(x)$ is decreasing for negative values of x. We read the graph from left to right: as x increases towards 0, the values of $f(x)$ decrease towards 0.

9. (a) (i) We have

$$\frac{f(2)-f(0)}{2-0} = \frac{16-2^2-(16-0)}{2} = -\frac{4}{2} = -2.$$

This means $f(x)$ decreases by an average of 2 units per unit change in x on the interval $0 \le x \le 2$.

(ii) We have

$$\frac{f(4)-f(2)}{4-2} = \frac{16-(4)^2-(16-2^2)}{2} = \frac{-16+4}{2} = -6.$$

This means $f(x)$ decreases by an average of 6 units per unit change in x on the interval $2 \le x \le 4$.

(iii) We have

$$\frac{f(4)-f(0)}{4-0} = \frac{16-(4)^2-(16-0)}{4} = -\frac{16}{4} = -4.$$

This means $f(x)$ decreases by an average of 4 units per unit change in x on the interval $0 \le x \le 4$.

(b) The graph of $f(x)$ is the solid curve in Figure 1.7. The secants corresponding to each rate of change are shown as dashed lines. The average rate of decrease is greatest on the interval $2 \le x \le 4$.

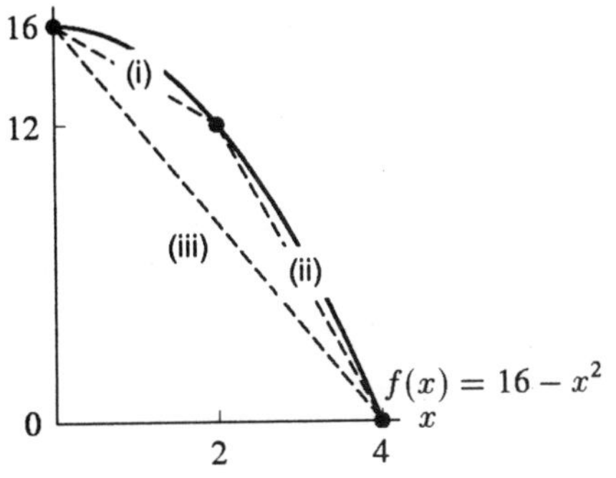

Figure 1.7

TABLE 1.1 *Carl Lewis' times at 10 meter intervals*

Time (sec)	Distance (meters)	$\Delta d/\Delta t$ (meters/sec)
0.00 to 1.94	0 to 10	5.15
1.94 to 2.96	10 to 20	9.80
2.96 to 3.91	20 to 30	10.53
3.91 to 4.78	30 to 40	11.49
4.78 to 5.64	40 to 50	11.63
5.64 to 6.50	50 to 60	11.63
6.50 to 7.36	60 to 70	11.63
7.36 to 8.22	70 to 80	11.63
8.22 to 9.07	80 to 90	11.76
9.07 to 9.93	90 to 100	11.63

13. (a) Table 1.1 shows the average rate of change of distance, commonly called the average speed or average velocity.

(b) He attained his maximum speed (11.76 meters/sec) between 80 and 90 meters. He does not appear to be running his fastest when he crossed the finish line.

Solutions for Chapter 1 Review

1. (a) Using the vertical line test, we can see that y is not a function of x.
   (b) To determine whether x is a function of y, we want to know if, for each value of y, there is a unique value of x associated with it. If we were to draw a horizontal line through the graph, representing one value of y, we could see that the line intersects the graph in more than one place. This tells us that there are many values of x corresponding to a value y, so this graph does not define a function.
   (c) If there were an interval on the x-axis for which y is a function of x, then there would be an interval for which each value of x would pass the vertical line test. The only place on the graph where that happens is in the interval shown in Figure 1.8.

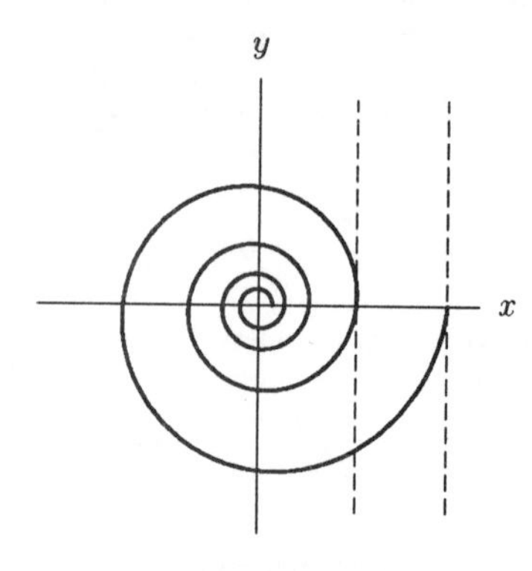

Figure 1.8

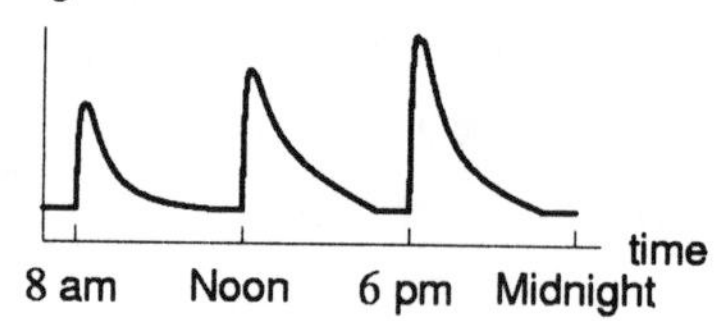

Figure 1.9

5. Figure 1.9 shows a possible graph of blood sugar level as a function of time over one day. Note that the actual curve is smooth, and does not have any sharp corners.

9. The original price is P. Inflation causes a 5% increase, giving

$$\text{Inflated price} = P + 0.05P = 1.05P.$$

Then there is a 10% decrease, giving

$$\begin{aligned}\text{Final price} &= 90\%(\text{Inflated price})\\ &= 0.9(1.05P)\\ &= 0.945P.\end{aligned}$$

13. No. This can be written as $y = kx^2$ with $k = 1$, but not as $y = kx$.

17. No. This can be written as $y = mx + b$ with $m = 1$ and $b = 2$, but it can not be written as $y = kx$.

21. The number of cookies is directly proportional to the number of cups of sugar, because an increased number of cups of sugar, s, increases the number of dozen cookies, n.

    Substituting the values we are given into the general formula $n = ks$, we get $2 = k(1.5)$, so $k = 1.33$, and the formula is

$$n = 1.33s.$$

    When $s = 2$, we have $n = 1.33(2) = 2.67$. Therefore 2 cups of sugar will yield 2.67 dozen or 32 cookies.

25. (a) We know that

$$\text{Time} = \frac{\text{Distance}}{\text{Rate}}, \quad \text{so} \quad T = \frac{D}{R}.$$

We solve for D:

$$\begin{aligned} D &= RT \\ &= (60 \text{ miles/hr})(20 \text{ hrs}) \\ &= 1200 \text{ miles.} \end{aligned}$$

(b) Since $D = 1200$, we have

$$T = \frac{1200}{R}.$$

The graph of this function is in Figure 1.10.

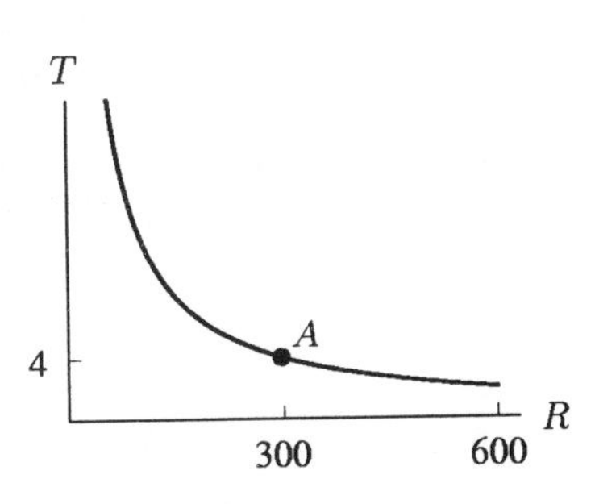

Figure 1.10

(c) Since

$$T = f(R) = \frac{1200}{R},$$

we have

$$f(300) = \frac{1200 \text{ miles}}{300 \text{ miles/hr}} = 4 \text{ hrs.}$$

In this context, $f(300)$ represents the number of hours required to travel 1200 miles at a rate of 300 miles/hr.

29. (a) Results are compiled in Table 1.2.

TABLE 1.2

n	1	2	3	4	5	6	7	8	9	10
$p(n)$	1	2	6	24	120	720	5,040	40,320	362,880	3,628,800

(b) Answers will vary.

33. (a) We know that 75% of David Letterman's 7 million person audience belongs to the nation's work force. Thus

$$\left(\begin{array}{c}\text{Number of people from the} \\ \text{work force in Dave's audience}\end{array}\right) = 75\% \text{ of 7 million} = 0.75 \cdot (7 \text{ million}) = 5.25 \text{ million.}$$

Thus the percentage of the work force in Dave's audience is

$$\left(\begin{array}{c}\text{\% of work force}\\ \text{in audience}\end{array}\right) = \left(\frac{\text{People from work force in audience}}{\text{Total work force}}\right)\cdot 100\%$$

$$= \left(\frac{5.25}{118}\right)\cdot 100\% = 4.45\%.$$

(b) Since 4.45% of the work force belongs to Dave's audience, David Letterman's audience must contribute 4.45% of the GDP. Since the GDP is estimated at \$6.325 trillion,

$$\left(\begin{array}{c}\text{Dave's audience's contribution}\\ \text{to the G.D.P.}\end{array}\right) = (0.0445)\cdot(6.325 \text{ trillion}) \approx 281 \text{ billion dollars.}$$

(c) Of the contributions by Dave's audience, 10% is estimated to be lost. Since the audience's total contribution is \$281 billion, the "Letterman Loss" is given by

$$\text{Letterman loss} = 0.1\cdot(281 \text{ billion dollars}) = \$28.1 \text{ billion.}$$

CHAPTER TWO

Solutions for Section 2.1

1. The function f could be linear because the value of x increases by $\Delta x = 5$ each time and $f(x)$ increases by $\Delta f(x) = 10$ each time. Assuming that any values of f not shown by the table follow this same pattern, the function f is linear.

   The function g is not linear even though $g(x)$ increases by $\Delta g(x) = 50$ each time. This is because the value of x does not increase by the same amount each time. The value of x increases from 0 to 100 to 300 to 600 taking steps that get larger each time.

   Similarly, the function h is not linear either even though the value of x increases by $\Delta x = 10$ each time. This is because $h(x)$ does not increase by the same amount each time. The value of $h(x)$ increases from 20 to 40 to 50 to 55 taking smaller steps each time.

   The function j could be linear because if the pattern continues for values of $j(x)$ that are not shown, we see that a one unit increase in x corresponds to a constant decrease of two units in y.

5. (a) Looking at the data from Table 2.11 and calculating the rate of change of area versus side length between various points, we see that the function is not linear. For example, the rate of change between the points $(0,0)$ and $(1,1)$ is

$$\frac{\Delta\text{area}}{\Delta\text{length}} = \frac{1-0}{1-0} = \frac{1}{1} = 1$$

while the rate of change between the points $(1,1)$ and $(2,4)$ is

$$\frac{\Delta\text{area}}{\Delta\text{length}} = \frac{4-1}{2-1} = \frac{3}{1} = 3.$$

The rates of change are different. The relationship is not linear. On the other hand, when we view the data from Table 2.11, we see that the rate of change of perimeter versus side length between any two points is always constant. Thus, that function is linear. For example, let's look at the pairs of points $(0,0)$, $(3,12)$; $(1,4)$, $(4,16)$ and $(2,8)$, $(5,20)$. For $(0,0)$, $(3,12)$ the rate of change is

$$\frac{\Delta\text{perimeter}}{\Delta\text{length}} = \frac{12-0}{3-0} = \frac{12}{3} = 4.$$

For $(1,4)$, $(4,16)$ the rate of change is

$$\frac{\Delta\text{perimeter}}{\Delta\text{length}} = \frac{16-4}{4-1} = \frac{12}{3} = 4.$$

For $(2,8)$, $(5,20)$ the rate of change is

$$\frac{\Delta\text{perimeter}}{\Delta\text{length}} = \frac{20-8}{5-2} = \frac{12}{3} = 4.$$

Check that using any two of the data points in Table 2.11 to calculate the rate of change gives a rate of change of 4. This function is linear.

(b) See Figures 2.1 and 2.2.

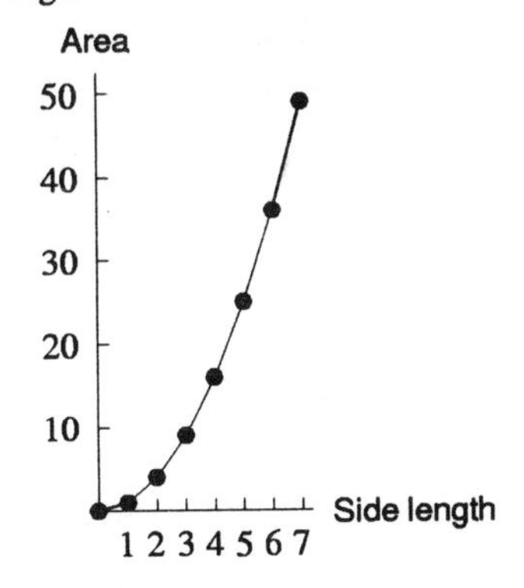

Figure 2.1: Area and side length

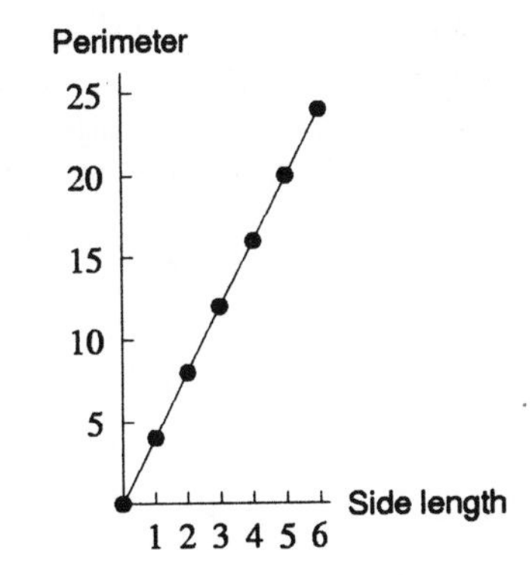

Figure 2.2: Perimeter and side length

(c) From part (a) we see that the rate of change of the function giving perimeter versus side length is 4. This tells us that for a given square, when we increase the length of each side by one unit, the length of the perimeter increases by four units.

9. The following table shows the population of the town as a function of the number of years since 1999.

TABLE 2.1

t	P
0	18,310
1	$18{,}310 + 58$
2	$18{,}310 + 58 + 58 = 18{,}310 + 2 \times 58$
3	$18{,}310 + 3 \times 58$
4	$18{,}310 + 4 \times 58$
$\cdots$	
t	$18{,}310 + t \times 58$

So, a formula is $P = 18{,}310 + 58t$.

13. (a) Since for each additional \$5000 spent the company will sell 20 more units, we have

$$m = \frac{\Delta y}{\Delta x} = \frac{20}{5000}.$$

Also, since 300 units will be sold even if no money is spent on advertising, the y–intercept, b, is 300. Our formula is

$$y = 300 + \frac{20}{5000}x = 300 + \frac{1}{250}x.$$

(b) If $x = \$25{,}000$, the number of units it sells will be

$$y = 300 + \frac{1}{250}(25000) = 300 + 100 = 400.$$

If $x = \$50{,}000$, the number of units it sells will be

$$y = 300 + \frac{1}{250}(50000) = 300 + 200 = 500.$$

(c) If $y = 700$, we need to solve for x:

$$\begin{aligned} 300 + \frac{1}{250}x &= 700 \\ \frac{1}{250}x &= 700 - 300 = 400 \\ x &= 250 \cdot 400 = 100{,}000. \end{aligned}$$

Thus, the firm would need to spend \$100,000 to sell 700 units.

(d) The slope is the change in the value of y, the number of units sold, for a given change in x, the amount of money spent on ads. Thus, an interpretation of the slope is that for each additional \$250 spent on ads, one additional unit is sold.

17. As Figure 2.3 shows, the graph is not visible in the window $10 \leq x \leq 10, -10 \leq y \leq 10$. The reason is that the graph of $y = 200x + 4$ is nearly vertical and almost coincides with the y-axis in this window. To see more clearly that the graph is not vertical, use a much larger y-range. For example, a window of $-10 \leq x \leq 10$, $-2000 \leq y \leq 2000$ gives a more informative graph. Alternatively, use a much smaller x-range; for example, try a window of $-0.1 \leq x \leq 0.1$, $-10 \leq y \leq 10$.

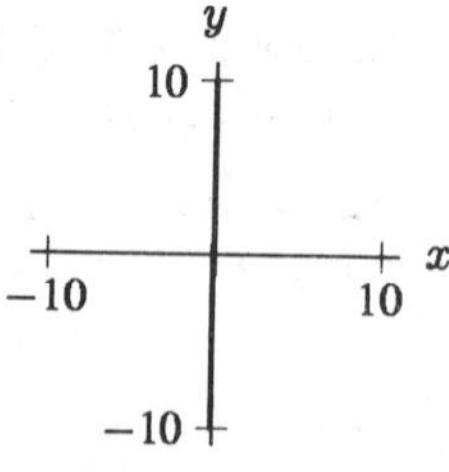

Figure 2.3

Solutions for Section 2.2

1. Rewriting in slope-intercept form:

$$\begin{aligned} 3x + 5y &= 20 \\ 5y &= 20 - 3x \\ y &= \frac{20}{5} - \frac{3x}{5} \\ y &= 4 - \frac{3}{5}x \end{aligned}$$

5. Rewriting in slope-intercept form:

$$\begin{aligned} 5x - 3y + 2 &= 0 \\ -3y &= -2 - 5x \\ y &= \frac{-2}{-3} - \frac{5}{-3}x \\ y &= \frac{2}{3} + \frac{5}{3}x \end{aligned}$$

9. Writing $y = 5$ as $y = 5 + 0x$ shows that $y = 5$ is the form $y = b + mx$ with $b = 5$ and $m = 0$.

13. Since we know the x-intercept and y-intercepts are $(3, 0)$ and $(0, -5)$ respectively, we can find the slope:

$$\text{slope} = m = \frac{-5-0}{0-3} = \frac{-5}{-3} = \frac{5}{3}.$$

We can then put the slope and y-intercept into the general equation for a line.

$$y = -5 + \frac{5}{3}x.$$

17. There are many possible answers. For example, when you buy something, the amount of sales tax depends on the sticker price of the item bought. Let's say Tax $= 0.05 \times$ Price. This means that the sales tax rate is 5%.

21. (a) This can be solved by finding the formula for the line through the two points $(30, 152.50)$ and $(60, 250)$. Here is an alternate approach. The membership fee will be the same for the 30-meal and 60-meal plans, while the fee for the meals themselves will depend on the number of meals. Thus,

$$\text{Total fee} = \text{Membership fee} + \text{Number of meals} \cdot \text{Price per meal}.$$

This gives us the two equations:

$$\begin{aligned} 152.50 &= M + 30 \cdot F \\ 250.00 &= M + 60 \cdot F, \end{aligned}$$

where M is the membership fee, and F is the fixed price per meal. Subtracting the first equation from the second and solving for F gives us:

$$\begin{aligned} 97.50 &= 30 \cdot F \\ F &= \frac{97.50}{30} = 3.25. \end{aligned}$$

Now that we know the fixed price per meal, we can use either of our original equations to solve for the membership fee, M:

$$\begin{aligned} 152.50 &= M + 30 \cdot 3.25 \\ M &= 152.50 - 97.50 = 55. \end{aligned}$$

Thus, the membership fee is \$55 and the price per meal is \$3.25.

(b) The cost of a meal plan is the membership fee plus n times the cost of a meal. Using our results from part (a):

$$C = 55 + 3.25 \cdot n.$$

(c) Using our formula for the cost of a meal plan:

$$C = 55 + 3.25 \cdot n = 55 + 3.25 \cdot 50 = \$217.50.$$

(d) Rewriting our expression for the cost of a meal plan:

$$\begin{aligned} 55 + 3.25 \cdot n &= C \\ 3.25 \cdot n &= C - 55 \\ n &= \frac{C - 55}{3.25}. \end{aligned}$$

(e) Given $C = \$300$ you can buy:

$$n = \frac{C - 55}{3.25} = \frac{300 - 55}{3.25} \approx 75.38.$$

Since the college is unlikely to sell you a fraction of a meal, we round this number down. Thus, 75 is the maximum number of meals you can buy for $300.

25. Point P is on the curve $y = x^2$ and so its coordinates are $(2, 2^2) = (2, 4)$. Since line l contains point P and has slope 4, its equation is

$$y = b + mx.$$

Using $P = (2, 4)$ and $m = 4$, we get

$$\begin{aligned} 4 &= b + 4(2) \\ 4 &= b + 8 \\ -4 &= b \end{aligned}$$

so,

$$y = -4 + 4x.$$

29. (a) Since q is linear, $q = b + mp$, where

$$\begin{aligned} m &= \frac{\Delta q}{\Delta p} = \frac{65 - 45}{1.10 - 1.50} \\ &= \frac{20}{-0.40} = -50 \text{ gallons/dollar}. \end{aligned}$$

Thus, $q = b - 50p$ and since $q = 65$ if $p = 1.10$,

$$\begin{aligned} 65 &= b - 50(1.10) \\ 65 &= b - 55 \\ b &= 65 + 55 = 120. \end{aligned}$$

So,

$$q = 120 - 50p.$$

(b) The slope is $m = -50$ gallons per dollar, which tells us that the quantity of gasoline demanded in one time period decreases by 50 gallons for each $1 increase in price.

(c) If $p = 0$ then $q = 120$, which means that if the price of gas were $0 per gallon, then the quantity demanded in one time period would be 120 gallons per month. This means if gas were free, a person would want 120 gallons. If $q = 0$ then $120 - 50p = 0$, so $120 = 50p$ and $p = \frac{120}{50} = 2.40$. This tells us that (according to the model), at a price of $2.40 per gallon there will be no demand for gasoline. In the real world, this is not likely.

Solutions for Section 2.3

1. One way to solve this system is by substitution. Solve the first equation for y:

$$\begin{aligned} 3x - y &= 17 \\ -y &= 17 - 3x \\ y &= 3x - 17. \end{aligned}$$

In the second equation, substitute the expression $3x - 17$ for y:

$$\begin{aligned} -2x - 3y &= -4 \\ -2x - 3(3x - 17) &= -4 \\ -2x - 9x + 51 &= -4 \\ -11x &= -4 - 51 = -55 \\ x &= \frac{-55}{-11} = 5. \end{aligned}$$

Since $x = 5$ and $y = 3x - 17$, we have

$$y = 3(5) - 17 = 15 - 17 = -2.$$

Thus, the solution to the system is $x = 5$ and $y = -2$.

Check your results by substituting the values into the second equation:

$$\begin{aligned} -2x - 3y &= -4 \\ \text{Substituting, we get } -2(5) - 3(-2) &= -4 \\ -10 + 6 &= -4 \\ -4 &= -4. \end{aligned}$$

5. (a)

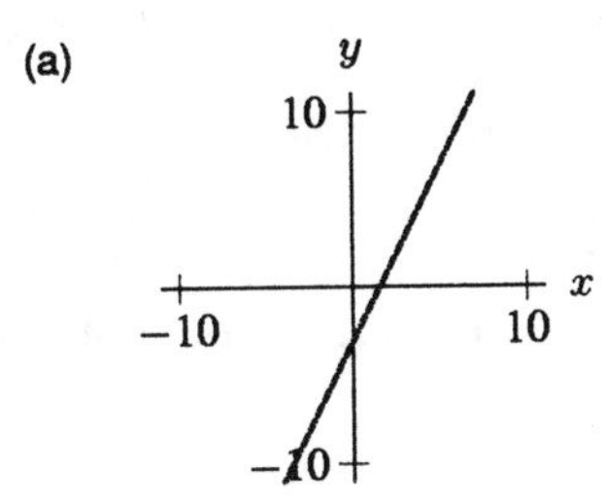

(b)

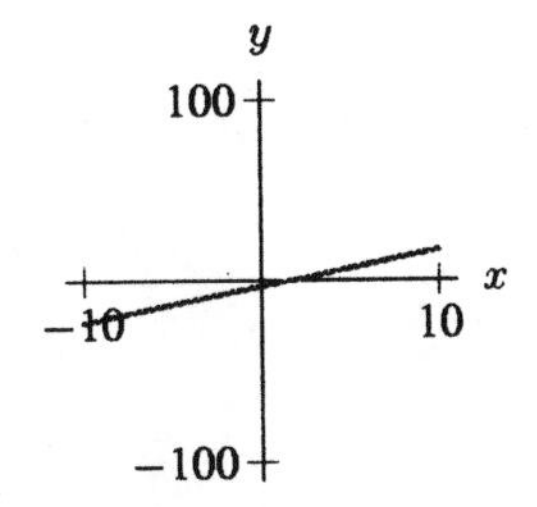

(c)

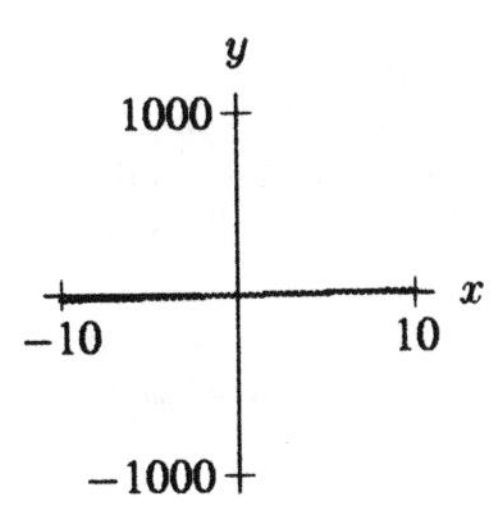

Figure 2.4

(d) If the width of the window remains constant and the height of the window increases, then the graph will appear less steep.

9. (a) The fixed cost is \$8000; \$200 is the unit cost.
   (b) The fixed cost is \$5000; \$200 is the unit cost.
   (c) The fixed cost is \$10,000; \$100 is the unit cost.
   (d) No fixed cost; \$50 is the unit cost.

13. (a) This line, being parallel to l, has the same slope. Since the slope of l is $-\frac{2}{3}$, the equation of this line is

$$y = b - \frac{2}{3}x.$$

To find b, we use the fact that $P = (6, 5)$ is on this line. This gives

$$\begin{aligned} 5 &= b - \frac{2}{3}(6) \\ 5 &= b - 4 \\ b &= 9. \end{aligned}$$

So the equation of the line is

$$y = 9 - \frac{2}{3}x.$$

(b) This line is perpendicular to line l, and so its slope is given by

$$m = \frac{-1}{-2/3} = \frac{3}{2}.$$

Therefore its equation is

$$y = b + \frac{3}{2}x.$$

We again use point P to find b:

$$\begin{aligned} 5 &= b + \frac{3}{2}(6) \\ 5 &= b + 9 \\ b &= -4. \end{aligned}$$

This gives

$$y = -4 + \frac{3}{2}x.$$

(c) Figure 2.5 gives a graph of line l together with point P and the two lines we have found.

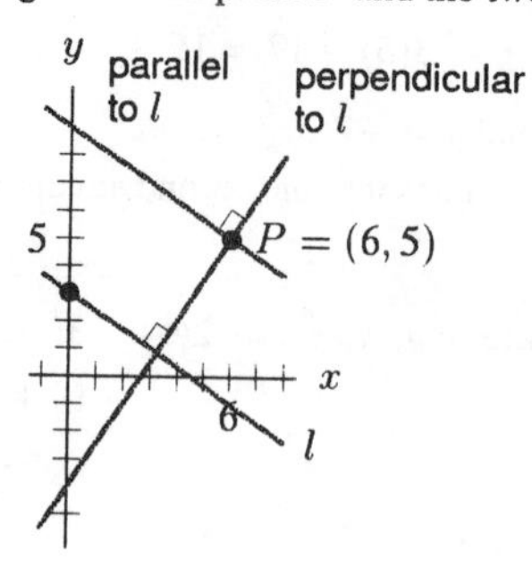

Figure 2.5: Line l and two lines through P, one parallel and one perpendicular to l

17. Point P lies on the two lines
$$y = 2x - 3.5 \quad \text{and} \quad y = -\frac{1}{2}x + 4.$$
One way to find P is to solve this system of equations simultaneously. Setting these two equations equal to each other and solving for x, we have
$$\begin{aligned} 2x - 3.5 &= -\frac{1}{2}x + 4 \\ 2x + \frac{1}{2}x &= 3.5 + 4 = 7.5 = \frac{15}{2} \\ \frac{5}{2}x &= \frac{15}{2} \\ x &= \frac{15}{2} \cdot \frac{2}{5} = 3. \end{aligned}$$
Since $x = 3$, we have
$$y = 2x - 3.5 = 2(3) - 3.5 = 6 - 3.5 = 2.5.$$
Thus, the coordinates of P are $(3, 2.5)$.

21. When $x = 1$, $y = \sqrt{1} = 1$, and when $x = 4$, $y = \sqrt{4} = 2$, so the points of intersection are $(1, 1)$ and $(4, 2)$. See Figure 2.6.

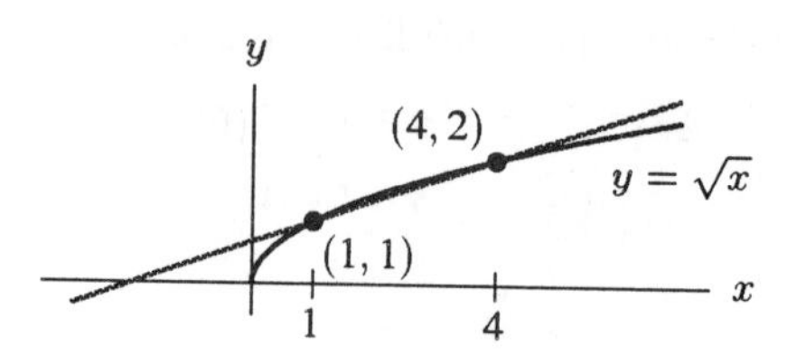

Figure 2.6

The line connecting $(1, 1)$ and $(4, 2)$ has slope $m = \frac{2-1}{4-1} = \frac{1}{3}$. To find the y-intercept, we can substitute one of the points, for example, $x = 1$, $y = 1$:
$$\begin{aligned} y &= \frac{1}{3}x + b \\ 1 &= \frac{1}{3}(1) + b \\ b &= \frac{2}{3} \end{aligned}$$
The equation of the line is $y = \frac{1}{3}x + \frac{2}{3}$. Now we'll solve the system
$$\begin{aligned} y &= \sqrt{x} \\ y &= \frac{1}{3}x + \frac{2}{3} \end{aligned}$$

by setting the equations equal to each other:

$$\sqrt{x} = \frac{1}{3}x + \frac{2}{3}.$$

Squaring both sides gives

$$\begin{aligned}
x &= \left(\frac{1}{3}x + \frac{2}{3}\right)^2 \\
x &= \frac{1}{9}x^2 + \frac{4}{9}x + \frac{4}{9} \\
\frac{1}{9}x^2 - \frac{5}{9}x + \frac{4}{9} &= 0 \\
x^2 - 5x + 4 &= 0 \quad \text{after multiplying both sides by 9} \\
(x-4)(x-1) &= 0 \\
x = 4 &\text{ or } x = 1
\end{aligned}$$

When $x = 4$, $y = \sqrt{4} = 2$, giving the point $(4, 2)$. When $x = 1$, $y = \sqrt{1} = 1$, giving the point $(1, 1)$. The results are consistent with the original problem

Solutions for Section 2.4

1. (a) Since the points lie on a line of positive slope, $r = 1$.
   (b) Although the points do not lie on a line, they are tending upward as x increases. So, there is a positive correlation and a reasonable guess is $r = 0.7$.
   (c) The points are scattered all over. There is neither an upward nor a downward trend, so there is probably no correlation between x and y, so $r = 0$.
   (d) These points are very close to lying on a line with negative slope, so the best correlation coefficient is $r = -0.98$.
   (e) Although these points are quite scattered, there is a downward slope, so $r = -0.25$ is probably a good answer.
   (f) These points are less scattered than those in part (e). The best answer here is $r = -0.5$.

5. (a) See Figure 2.7.
   (b) The scatterplot suggests that as IQ increases, the number of hours of TV viewing decreases. The points, though, are not close to being on a line, so a reasonable guess is $r \approx -0.5$.
   (c) A calculator gives the regression equation $y = 27.5139 - 0.1674x$ with $r = -0.5389$.

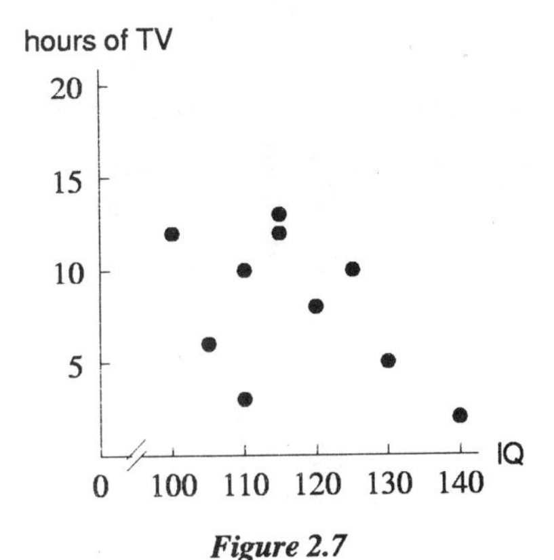

Figure 2.7

Solutions for Chapter 2 Review

1. We can write the equation in slope-intercept form

$$\begin{aligned} 3x + 5y &= 6 \\ 5y &= 6 - 3x \\ y &= \frac{6}{5} - \frac{3}{5}x. \end{aligned}$$

The slope is $\frac{-3}{5}$. Lines parallel to this line all have slope $\frac{-3}{5}$. Since the line passes through $(0, 6)$, its y-intercept is equal to 6. So $y = 6 - \frac{3}{5}x$.

5. (a) is (V), because slope is negative, vertical intercept is 0
   (b) is (VI), because slope and vertical intercept are both positive
   (c) is (I), because slope is negative, vertical intercept is positive
   (d) is (IV), because slope is positive, vertical intercept is negative
   (e) is (III), because slope and vertical intercept are both negative
   (f) is (II), because slope is positive, vertical intercept is 0

9. Since you are moving in a straight line away from Pittsburgh, your total distance is the initial distance, 60 miles, plus the additional miles covered. In each hour, you will travel fifty miles as shown in Figure 2.8.

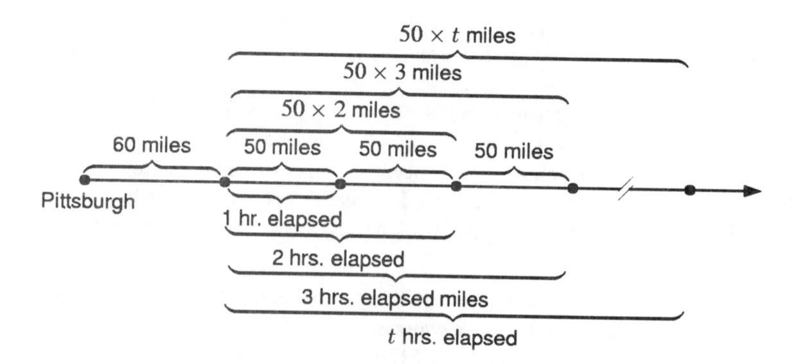

Figure 2.8

So, the total distance from Pittsburgh can be expressed as $d = 60 + 50t$.

13. (a) To stretch the spring a greater length, one would expect to need a greater force. Therefore, as x increases we find that $F(x)$ increases. Therefore, $F(x)$ is an increasing function of x.
    (b) Substituting 1.9 for x and 2.36 for $F(x)$, we get

$$\begin{aligned} 2.36 &= k(1.9) \\ \frac{2.36}{1.9} &= k \\ k &\approx 1.242 \\ F(x) &= 1.242x. \end{aligned}$$

    (c) Setting $x = 3$, we get $F(3) = 1.242(3) = 3.726$. Therefore, a force of approximately 3.7 pounds is required to hold the spring stretched 3 inches.

17. Let us write the equation for the diameter $d(g)$ as follows:

$$d(g) = b + mg$$

where g is the gauge number (and in our case the independent variable), m is the slope of the function and b is the d-intercept. First find the slope, m, by using the data points $(2, 0.2656)$ and $(8, 0.1719)$:

$$\begin{aligned} m &= \frac{d(8) - d(2)}{8 - 2} \\ &= \frac{0.1719 - 0.2656}{8 - 2} = \frac{-.0937}{6} \approx -0.01562. \end{aligned}$$

We will use 5 decimal places. Thus $d(g) = b + (-0.01562)g$. Substituting the point $(2, 0.2656)$ in this equation and solving for b, gives

$$\begin{aligned} 0.2656 &= b + (-0.01562)(2) \\ 0.2656 &= b + (-0.03124) \\ \text{and} \quad b &= 0.29684. \end{aligned}$$

Thus,

$$d(g) = 0.29684 + (-0.01562)g.$$

So,

$$\begin{aligned} d(12.5) &= (-0.01562)(12.5) + 0.29684 \\ &= -0.19525 + 0.29684 = 0.10159 \end{aligned}$$

and

$$d(0) = (-0.01562)(0) + 0.29684 = 0.29684.$$

Thus, gauge 12.5 corresponds to a thickness of 0.1016 inches, while gauge 0 corresponds to a thickness of 0.2968 inches. We know that gauge numbers are no longer sensible when they correspond to a negative or zero thickness, thus we must solve

$$d(g) > 0.$$

Solving, we get

$$\begin{aligned} d(g) &> 0 \\ (-0.01562)g + 0.29684 &> 0 \\ (-0.01562)g &> -0.29684 \\ g &< \frac{-0.29684}{-0.01562} \approx 19 \quad \text{(since we divided by a negative number, we must flip the inequality sign).} \end{aligned}$$

Thus, the gauge number only makes sense for values less than 19.

21. (a) When the price of the product went from \$3 to \$4, the demand for the product went down by 200 units. Since we are assuming that this relationship is linear, we know that the demand will drop by another 200 units when the price increases another dollar, to \$5. When $p = 5$, $D = 300 - 200 = 100$. So, when the price for each unit is \$5, consumers will only buy 100 units a week.

(b) The slope, m, of a linear equation is given by

$$m = \frac{\text{change in dependent variable}}{\text{change in independent variable}} = \frac{\Delta D}{\Delta P}.$$

Since quantity demanded depends on price, quantity demanded is the dependent variable and price is the independent variable. We know that when the price changes by $1, the quantity demand changes by −200 units. That is, the quantity demanded goes down by 200 units. Thus,

$$m = \frac{-200}{1}.$$

Since the relationship is linear, we know that its formula is of the form

$$D = b + mp.$$

We know that $m = -200$, so

$$D = b - 200p.$$

We can find b by using the fact that when $p = 3$ then $D = 500$ or by using the fact that if $p = 4$ then $D = 300$ (it doesn't matter which). Using $p = 3$ and $D = 500$, we get

$$\begin{aligned} D &= b - 200p \\ 500 &= b - 200(3) \\ 500 &= b - 600 \\ 1100 &= b. \end{aligned}$$

Thus, $D = 1100 - 200p$.

(c) We know that $D = 1100 - 200p$ and $D = 50$, so

$$\begin{aligned} 50 &= 1100 - 200p \\ -1050 &= -200p \\ 5.25 &= p. \end{aligned}$$

At a price of $5.25, the demand would be only 50 units.

(d) The slope is −200, which means that the demand goes down by 200 units when the price goes up by $1.

(e) The demand is 1100 when the price is 0. This means that even if you were giving this product away, people would only want 1100 units of it per week. When the price is $5.50, the demand is zero. This means that at or above a unit price of $5.50, the company cannot sell this product.

25. (a)

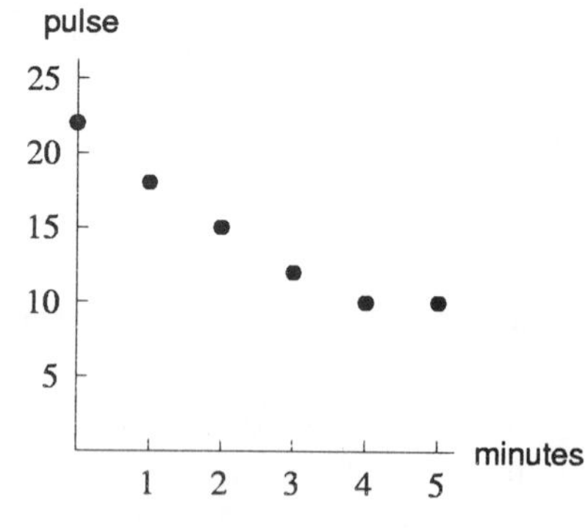

Figure 2.9

(b) For $0 \leq t \leq 4$, the pulse values nearly lie on a straight line.
(c) The correlation is close to $r = -1$ for time less than 4 minutes. After 4 minutes, the pulse rate reaches its normal, constant level, and there would be no correlation.

29. In Figure 2.10 the decision to spend all c dollars of your money on apples is represented by the x-intercept; the decision to spend it all on bananas is represented the y-intercept. If we decide to spend all of our money on either all apples or all bananas, and bananas are cheaper, then we would be able to purchase more bananas than apples for our c dollars. So, we want the line for which the y-intercept is greater than the x-intercept. If we look at line l_1 we see that the y-intercept is greater than 10 and the x–intercept is less than 10. Thus, l_1 represents the case where we can buy more bananas than apples, so apples must be more expensive than bananas.

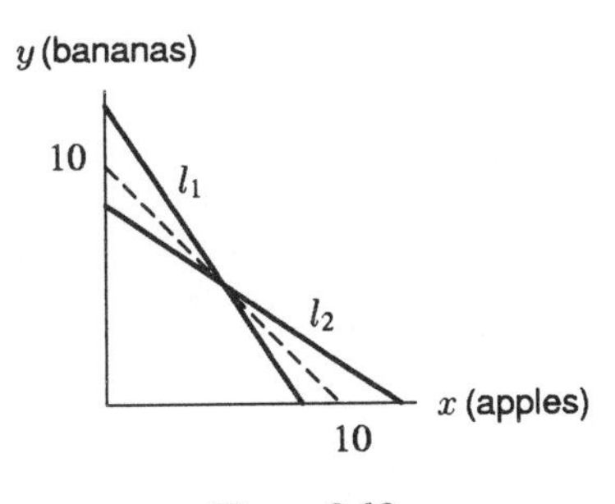

Figure 2.10

CHAPTER THREE

Solutions for Section 3.1

1. The input, t, is the number of months since January 1, and the output, F, is the number of foxes. The expression $g(9)$ represents the number of foxes in the park on October 1. Table 1.4 on page 5 of the text gives $F = 100$ when $t = 9$. Thus, $g(9) = 100$. On October 1, there were 100 foxes in the park.

5. (a) Substituting $t = 0$ gives $f(0) = 0^2 - 4 = -4$.
   (b) Setting $f(t) = 0$ and solving gives $t^2 - 4 = 0$, so $t^2 = 4$, so $t = \pm 2$.

9. (a)

x	-2	-1	0	1	2	3
$h(x)$	0	9	8	3	0	6

   (b) $h(3) = 6$, while $h(1) = 3$. Thus, $h(3) - h(1) = 6 - 3 = 3$.
   (c) $h(2) = 0$, and $h(0) = 8$. Thus, $h(2) - h(0) = 0 - 8 = -8$.
   (d) From the table, we see that $h(0) = 8$. Thus, $2h(0) = 2(8) = 16$.
   (e) From the table, we see that $h(1) = 3$. Thus, $h(1) + 3 = 3 + 3 = 6$.

13. (a) Substituting into $h(t) = -16t^2 + 64t$, we get

$$h(1) = -16(1)^2 + 64(1) = 48$$
$$h(3) = -16(3)^2 + 64(3) = 48$$

   Thus the height of the ball is 48 feet after 1 second and after 3 seconds.

   (b) The graph of $h(t)$ is in Figure 3.1. The ball is on the ground when $h(t) = 0$. From the graph we see that this occurs at $t = 0$ and $t = 4$. The ball leaves the ground when $t = 0$ and hits the ground at $t = 4$ or after 4 seconds. From the graph we see that the maximum height is 64 ft.

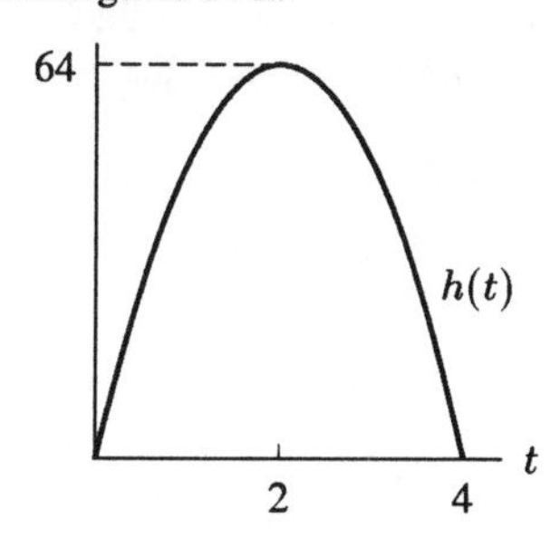

Figure 3.1

17. (a)

n	1	2	3	4	5
$s(n)$	1	3	6	10	15

   (b) Substituting into the formula for $s(n)$, we have

$$s(1) = \frac{1(1+1)}{2} = \frac{1 \cdot 2}{2} = 1$$
$$s(2) = \frac{2(2+1)}{2} = \frac{2 \cdot 3}{2} = 3$$
$$s(3) = \frac{3(3+1)}{2} = \frac{3 \cdot 4}{2} = 6$$
$$s(4) = \frac{4(4+1)}{2} = \frac{4 \cdot 5}{2} = 10$$
$$s(5) = \frac{5(5+1)}{2} = \frac{5 \cdot 6}{2} = 15.$$

(c) To find out how many pins are needed for a 100 row arrangement, we evaluate $s(100)$:

$$s(100) = \frac{100 \cdot 101}{2} = 5050.$$

So 5050 pins are needed.

Solutions for Section 3.2

1. (a) (i) To evaluate $f(x)$ for $x = 6$, we find from the table the value of $f(x)$ corresponding to an x-value of 6. In this case, the corresponding value is 248. Thus, $f(x)$ at $x = 6$ is 248.
(ii) $f(5)$ equals the value of $f(x)$ corresponding to $x = 5$, or 145. $f(5) - 3 = 145 - 3 = 142$.
(iii) $f(5-3)$ is the same thing as $f(2)$, which is the value of $f(x)$ corresponding to $x = 2$. Since $f(5-3) = f(2)$, and $f(2) = 4$, $f(5-3) = 4$.
(iv) $g(x)+6$ for $x = 2$ equals $g(2)+6$. $g(2)$ is the value of $g(x)$ corresponding to an x-value of 2, thus $g(2) = 6$. $g(2) + 6 = 6 + 6 = 12$.
(v) $g(x + 6)$ for $x = 2$ equals $g(2 + 6) = g(8)$. Looking at the table in the problem, we see that $g(8) = 378$. Thus, $g(x + 6)$ for $x = 2$ equals 378.
(vi) $g(x)$ for $x = 0$ equals $g(0) = -6$. $3 \cdot (g(0)) = 3 \cdot (-6) = -18$.
(vii) $f(3x)$ for $x = 2$ equals $f(3 \cdot 2) = f(6)$. From part (a), we know that $f(6) = 248$; thus, $f(3x)$ for $x = 2$ equals 248.
(viii) $f(x) - f(2)$ for $x = 8$ equals $f(8) - f(2)$. $f(8) = 574$ and $f(2) = 4$, so $f(8) - f(2) = 574 - 4 = 570$.
(ix) $g(x + 1) - g(x)$ for $x = 1$ equals $g(1 + 1) - g(1) = g(2) - g(1)$. $g(2) = 6$ and $g(1) = -7$, so $g(2) - g(1) = 6 - (-7) = 6 + 7 = 13$.

(b) (i) To find x such that $g(x) = 6$, we look for the entry in the table at which $g(x) = 6$ and then see what the corresponding x-value is. In this case, it is 2. Thus, $g(x) = 6$ for $x = 2$.
(ii) We use the same principle as that in part (i): $f(x) = 574$ when $x = 8$.
(iii) Again, this is just like part (i): $g(x) = 281$ when $x = 7$.

(c) Solving $x^3 + x^2 + x - 10 = 7x^2 - 8x - 6$ involves finding those values of x for which both sides of the equation are equal, or where $f(x) = g(x)$. Looking at the table, we see that $f(x) = g(x) = -7$ for $x = 1$, and $f(x) = g(x) = 74$ for $x = 4$.

5. (a) To find a point on the graph of $h(x)$ whose x-coordinate is 5, we substitute 5 for x in the formula for $h(x)$. $h(5) = \sqrt{5+4} = \sqrt{9} = 3$. Thus, the point $(5, 3)$ is on the graph of $h(x)$.

(b) Here we want to find a value of x such that $h(x) = 5$. We set $h(x) = 5$ to obtain

$$\begin{aligned} \sqrt{x+4} &= 5 \\ x + 4 &= 25 \\ x &= 21. \end{aligned}$$

Thus, $h(21) = 5$, and the point $(21, 5)$ is on the graph of $h(x)$.

(c) Figure 3.2 shows the desired graph. The point in part (a) is $(5, h(5))$, or $(5, 3)$. This point is labeled A in Figure 3.2. The point in part (b) is $(21, 5)$. This point is labeled B in Figure 3.2.

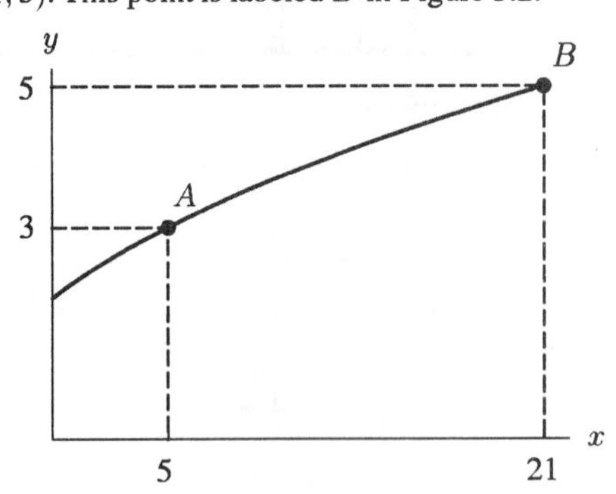

Figure 3.2

(d) If $p = 2$, then $h(p+1) - h(p) = h(2+1) - h(2) = h(3) - h(2)$. But $h(3) = \sqrt{3+4}$, while $h(2) = \sqrt{2+4}$, thus, $h(p+1) - h(p)$ for $p = 2$ equals $h(3) - h(2) = \sqrt{7} - \sqrt{6} \approx 0.1963$.

9. (a) (iii) The number of gallons needed to cover the house is $f(A)$; two more gallons will be $f(A) + 2$.
(b) (i) To cover the house twice, you need $f(A) + f(A) = 2f(A)$.
(c) (ii) The sign is an extra 2 ft^2 so we need to cover the area $A + 2$. Since $f(A)$ is the number of gallons needed to cover A square feet, $f(A+2)$ is the number of gallons needed to cover $A + 2$ square feet.

13. I is (b)
II is (d)
III is (c)
IV is (h)

17. (a) This is the fare for a ride of 3.5 miles. $C(3.5) \approx \$6.25$.
(b) This is the number of miles you can travel for \$3.50. Between 1 and 2 miles the increase in cost is \$1.50. Setting up a proportion we have:

$$\frac{1 \text{ additional mile}}{\$1.50 \text{ additional fare}} = \frac{x \text{ additional miles}}{\$3.50 - \$2.50 \text{ additional fare}}$$

and $x = 0.67$ miles. Therefore

$$C^{-1}(\$3.5) \approx 1.67.$$

Solutions for Section 3.3

1. The domain is all possible input values, namely $t = 1, 2, 3, \ldots, 12$.

5. (a) We must have $4 - x^2 \geq 0$, that is, $x^2 \leq 4$, so the domain of $f(x)$ is $-2 \leq x \leq 2$.
(b)

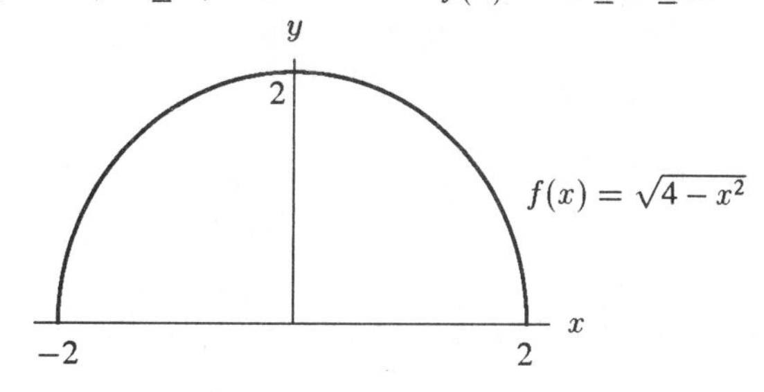

Figure 3.3

(c) Since $0 \leq \sqrt{4 - x^2} \leq \sqrt{4} = 2$, the range of $f(x)$ is $0 \leq y \leq 2$. This can also be seen from Figure 3.3.

9. The graph of $y = 1/x^2$ is given in Figure 3.4. The domain is all real numbers x, $x \neq 0$; the range is all $y > 0$.

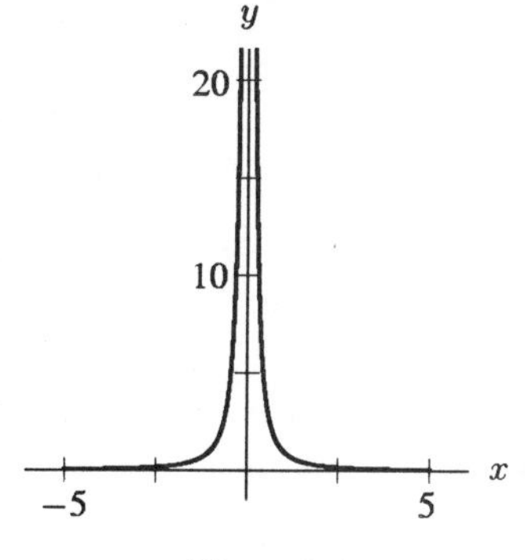

Figure 3.4

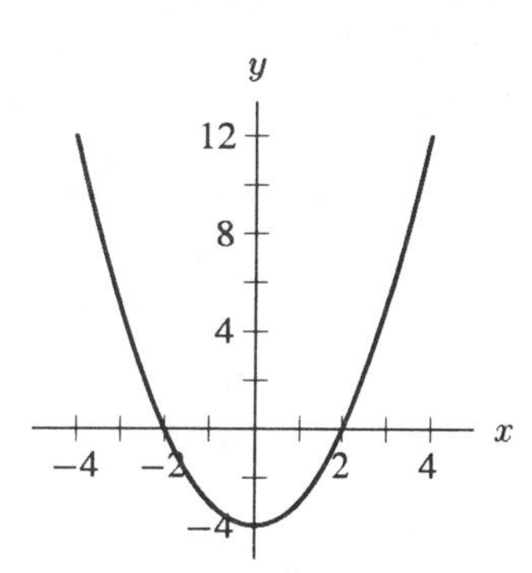

Figure 3.5

13. The graph of $y = x^2 - 4$ is given in Figure 3.5. The domain is all real x; the range is all $y \geq -4$.

17. The graph of $y = (x-4)^3$ is given in Figure 3.6. The domain is all real x; the range is all real y.

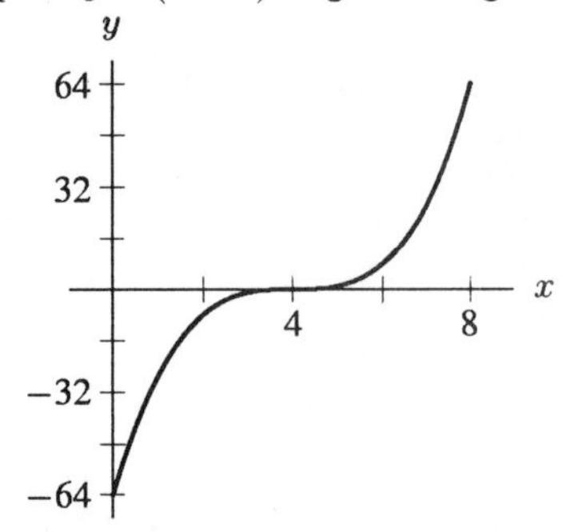

Figure 3.6

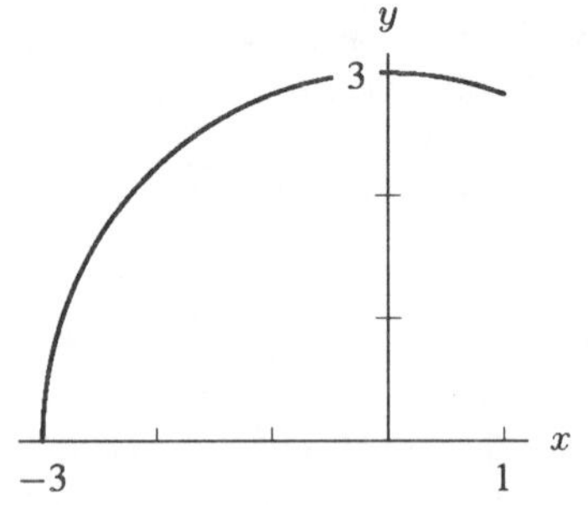

Figure 3.7

21. The graph of $y = \sqrt{9-x^2}$ for $-3 \le x \le 1$ is shown in Figure 3.7. From the graph, we see that $y = 0$ at $x = -3$, and that y increases to a maximum value of 3 at $x = 0$, and then decreases to a value of $y = \sqrt{9-1^2} \approx 2.83$ or $= 2\sqrt{2}$ at $x = 1$. Thus, on the domain $-3 \le x \le 1$, the range is $0 \le y \le 3$.

25. (a) We see that the 6th listing has a last digit of 8. Thus, $f(6) = 8$.
    (b) The domain of the telephone directory function is $n = 1, 2, 3, \ldots, N$, where N is the total number of listings in the directory. We could find the value of N by counting the number of listings in the phone book.
    (c) The range of this function is $d = 0, 1, 2, \ldots, 9$, because the last digit of any listing must be one of these numbers.

Solutions for Section 3.4

1. (a) Yes, because every value of x is associated with exactly one value of y.
   (b) No, because some values of y are associated with more than one value of x.
   (c) $y = 1, 2, 3, 4$.

5. The graph of $f(x) = \begin{cases} x+4, & x \le -2 \\ 2, & -2 < x < 2 \\ 4-x, & x \ge 2 \end{cases}$ is shown in Figure 3.8.

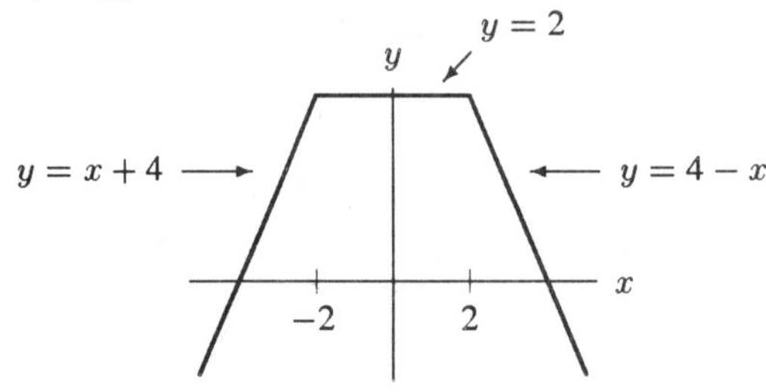

Figure 3.8

9. (a) The depth of the driveway is 1 foot or 1/3 of a yard. The volume of the driveway is the product of the three dimensions, length, width and depth. So,

$$\text{Volume of gravel needed} = \text{Length} \cdot \text{Width} \cdot \text{Depth} = (L)(6)(1/3) = 2L.$$

Since he buys 10 cubic yards more than needed,

$$n(L) = 2L + 10.$$

   (b) The length of a driveway is not less than 5 yards, so the domain of n is all real numbers greater than or equal to 5. The contractor can buy only 1 cubic yd at a time so the range only contains integers. The smallest value of the range occurs for the shortest driveway, when $L = 5$. If $L = 5$, then $n(5) = 2(5) + 10 = 20$. Although very long driveways are highly unlikely, there is no upper limit on L, so no upper limit on $n(L)$. Therefore the range is all integers greater than or equal to 20. See Figure 3.9.

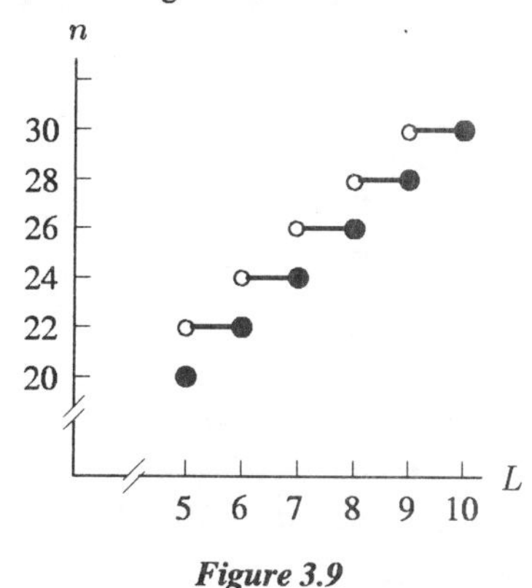

Figure 3.9

(c) If $n(L) = 2L + 10$ was not intended to represent a quantity of gravel, then the domain and range of n would be all real numbers.

13. (a) Each signature printed costs \$0.14, and in a book of p pages, there are at least $p/16$ signatures. In a book of 128 pages, there are

$$\frac{128}{16} = 8 \text{ signatures,}$$

$$\text{Cost for 128 pages } = 0.14(8) = \$1.12.$$

A book of 129 pages requires 9 signatures, although the ninth signature is used to print only 1 page. Therefore,

$$\text{Cost for 129 pages } = \$0.14(9) = \$1.26.$$

To find the cost of p pages, we first find the number of signatures. If p is divisible by 16, then the number of signatures is $p/16$ and the cost is

$$C(p) = 0.14\left(\frac{p}{16}\right).$$

If p is not divisible by 16, the number of signatures is $p/16$ rounded up to the next highest integer and the cost is 0.14 times that number. In this case, it is hard to write a formula for $C(p)$ without a symbol for "rounding up."

(b) The number of pages, p, is greater than zero. Although it is possible to have a page which is only half filled, we do not say that a book has 124 1/2 pages, so p must be an integer. Therefore, the domain of $C(p)$ is $p > 0$, p an integer. Because the cost of a book increases by multiples of \$0.14 (the cost of one signature), the range of $C(p)$ is $C > 0$, C an integer multiple of \$0.14,

(c) For 1 to 16 pages, the cost is \$0.14, because only 1 signature is required. For 17 to 32 pages, the cost is \$0.28, because 2 signatures are required. These data are continued in Table 3.1 for $0 \le p \le 128$, and they are plotted in Figure 3.10. A closed circle represents a point included on the graph, and an open circle indicates a point excluded from the graph. The unbroken lines in Figure 3.10 suggest, erroneously, that *fractions* of pages can be printed. It would be more accurate to draw each step as 16 separate dots instead of as an unbroken line.

TABLE 3.1 *The cost C for printing a book of p pages*

p, pages	C, dollars
1-16	0.14
17-32	0.28
33-48	0.42
49-64	0.56
65-80	0.70
81-96	0.84
97-112	0.98
113-128	1.12

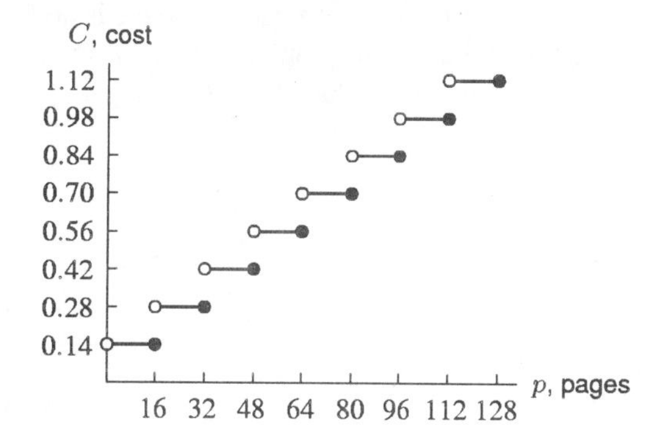

Figure 3.10: Graph of the cost C for printing a book of p pages

Solutions for Chapter 3 Review

1.

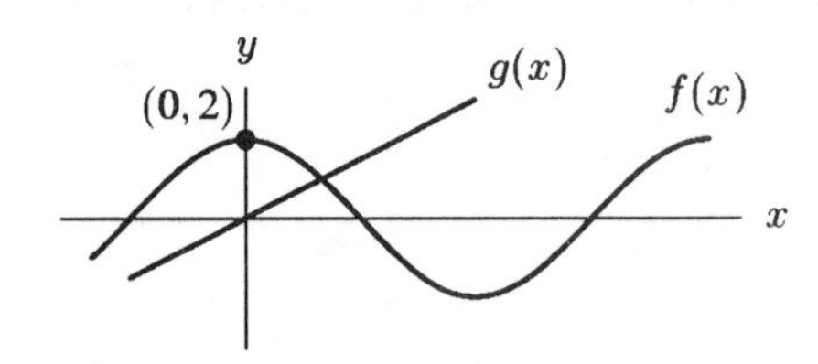

5. (a) $g(100) = 100\sqrt{100} + 100 \cdot 100 = 100 \cdot 10 + 100 \cdot 100 = 11,000$
   (b) $g(4/25) = 4/25 \cdot \sqrt{4/25} + 100 \cdot 4/25 = 4/25 \cdot 2/5 + 16 = 8/125 + 16 = 16.064$
   (c) $g(1.21 \cdot 10^4) = g(12100) = (12100)\sqrt{12100} + 100 \cdot (12100) = 2,541,000$

9. (a) A table of values for $y = 1/x$ follows:

x	-5	-4	-3	-2	-1	$-\frac{3}{4}$	$-\frac{1}{2}$	$-\frac{1}{4}$
y	$-\frac{1}{5}$	$-\frac{1}{4}$	$-\frac{1}{3}$	$-\frac{1}{2}$	-1	$-\frac{4}{3}$	-2	-4

x	0	$\frac{1}{4}$	$\frac{1}{2}$	$\frac{3}{4}$	1	2	3	4	5
y	undefined	4	2	$\frac{4}{3}$	1	$\frac{1}{2}$	$\frac{1}{3}$	$\frac{1}{4}$	$\frac{1}{5}$

   (b) The y-values in the table range from -4 to 4, but values of x close to 0 give y-values of very large magnitude.

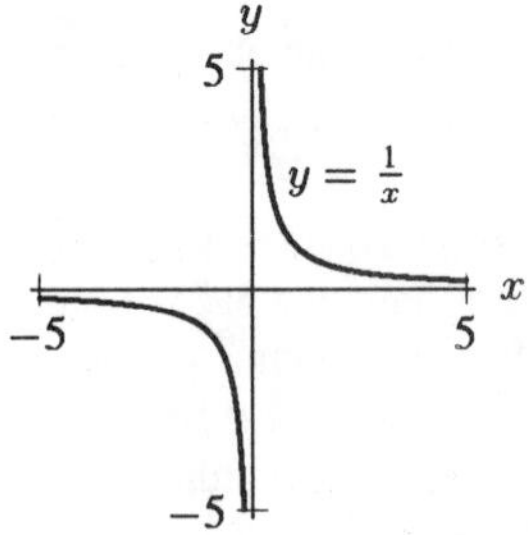

Figure 3.11

   (c) Domain is all real numbers except 0.
   Range is all real numbers except 0.
   (d) Decreasing: $-\infty < x < 0$ and $0 < x < \infty$.

13. (a) If $x = a$ is not in the domain of f there is no point on the graph with x-coordinate a. For example, there are no points on the graph of the function in Figure 3.12 with x-coordinates greater than 2. Therefore, $x = a$ is not in the domain of f for any $a > 2$.

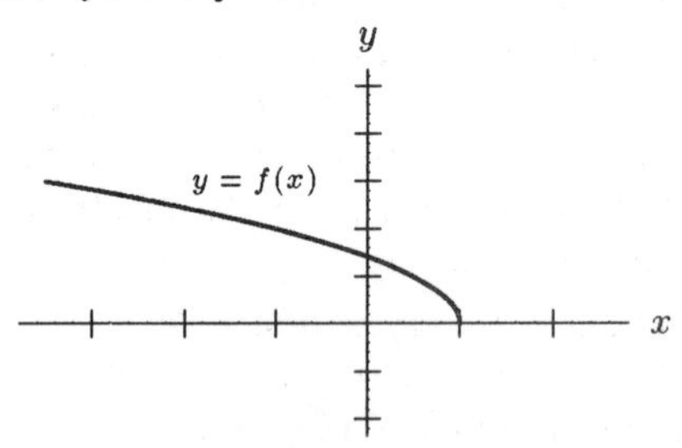

Figure 3.12

(b) If $x = a$ is not in the domain of f the formula is undefined for $x = a$. For example, if $f(x) = 1/(x-3)$, $f(3)$ is undefined, so 3 is not in the domain of f.

17. We know that $x \geq 4$, for otherwise $\sqrt{x-4}$ would be undefined. We also know that $4 - \sqrt{x-4}$ must not be negative. Thus we have

$$\begin{aligned} 4 - \sqrt{x-4} &\geq 0 \\ 4 &\geq \sqrt{x-4} \\ 4^2 &\geq \left(\sqrt{x-4}\right)^2 \\ 16 &\geq x - 4 \\ 20 &\geq x. \end{aligned}$$

Thus, the domain of $r(x)$ is $4 \leq x \leq 20$.

We use a computer or graphing calculator to find the range of $r(x)$. Graphing over the domain of $r(x)$ gives Figure 3.13.

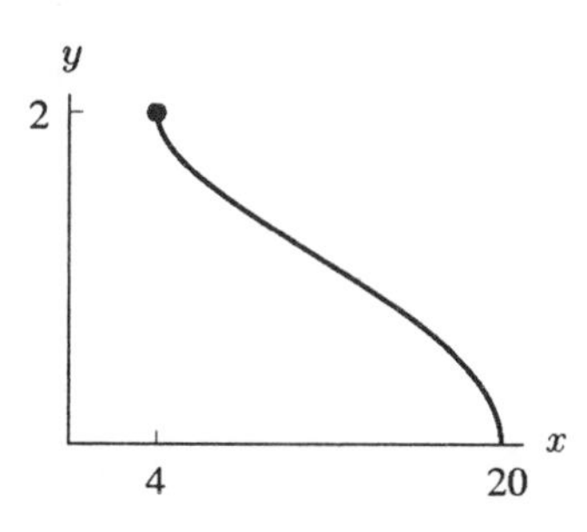

Figure 3.13

Because $r(x)$ is a decreasing function, we know that the maximum value of $r(x)$ occurs at the left end point of the domain, $r(4) = \sqrt{4 - \sqrt{4-4}} = \sqrt{4-0} = 2$, and the minimum value of $r(x)$ occurs at the right end point, $r(20) = \sqrt{4 - \sqrt{20-4}} = \sqrt{4 - \sqrt{16}} = \sqrt{4-4} = 0$. The range of $r(x)$ is thus $0 \leq r(x) \leq 2$.

21. (a) To evaluate $f(2)$, we determine which value of I corresponds to $w = 2$. Looking at the graph, we see that $I \approx 7$ when $w = 2$. This means that $\approx$ 7000 people were infected two weeks after the epidemic began.
(b) The height of the epidemic occurred when the largest number of people were infected. To find this, we look on the graph to find the largest value of I, which seems to be approximately 8.5, or 8500 people. This seems to have occurred when $w = 4$, or four weeks after the epidemic began. We can say that at the height of the epidemic, at $w = 4$, $f(4) = 8.5$.
(c) To solve $f(x) = 4.5$, we must find the value of w for which $I = 4.5$, or 4500 people were infected. We see from the graph that there are actually two values of w at which $I = 4.5$, namely $w \approx 1$ and $w \approx 10$. This means that 4500 people were infected after the first week when the epidemic was on the rise, and that after the tenth week, when the epidemic was slowing, 4500 people remained infected.
(d) We are looking for all the values of w for which $f(w) \geq 6$. Looking at the graph, this seems to happen for all values of $w \geq 1.5$ and $w \leq 7.5$. This means that more than 6000 people were infected starting in the middle of the second week and lasting until the middle of the eighth week, after which time the number of infected people fell below 6000.

25. (a) $f(2) = 1.50$ means that 2 kilograms of apples cost \$1.50.
(b) $f(0.1) = 0.75$ means that 0.1 kilogram of apples cost \$0.75 or 75¢.
(c) $f^{-1}(3) = 4$ means that \$3 buys 4 kilograms of apples.
(d) $f^{-1}(1.5) = 2$ means that \$1.50 buys 2 kilograms of apples.

29. (a) (i) From the table, $N(150) = 6$. When 150 students enroll, there are 6 sections.
(ii) Since $N(75) = 4$ and $N(100) = 5$, and 80 is between 75 and 100 students, we choose the higher value for $N(s)$. So $N(80) = 5$. When 80 students enroll, there are 5 sections.
(iii) The quantity $N(55.5)$ is not defined, since 55.5 is not a possible number of students.
(b) (i) The table gives $N(s) = 4$ sections for $s = 75$ and $s = 50$. For any integer between those in the table, the section number is the higher value. Therefore, for $50 \leq s \leq 75$, we have $N(s) = 4$ sections. We do not know what happens if $s < 50$.
(ii) First evaluate $N(125) = 5$. So we solve the equation $N(s) = 5$ for s. There are 5 sections when enrollment is between 76 and 125 students.

CHAPTER FOUR

Solutions for Section 4.1

1. The percent of change is given by

$$\text{Percent of change} = \frac{\text{Amount of change}}{\text{Old amount}} \cdot 100\%.$$

So in these two cases,

$$\text{Percent of change from 10 to 12} = \frac{12-10}{10} \cdot 100\% = 20\%$$

$$\text{Percent of change from 100 to 102} = \frac{102-100}{100} \cdot 100\% = 2\%$$

5. To find a formula for $f(n)$, we start with the number of people infected in 1990, namely P_0. In 1991, only 80% as many people, or $0.8P_0$, were infected. In 1992, again only 80% as many people were infected, which means that 80% of $0.8P_0$ people, or $0.8(0.8P_0)$ people, were infected. Continuing this line of reasoning, we can write

$$f(0) = P_0$$

$$f(1) = \underbrace{(0.80)}_{\text{one 20\% reduction}} P_0 = (0.8)^1 P_0$$

$$f(2) = \underbrace{(0.80)(0.80)}_{\text{two 20\% reductions}} P_0 = (0.8)^2 P_0$$

$$f(3) = \underbrace{(0.80)(0.80)(0.80)}_{\text{three 20\% reductions}} P_0 = (0.8)^3 P_0,$$

and so on, so that n years after 1990 we have

$$f(n) = \underbrace{(0.80)(0.80)\cdots(0.80)}_{n \text{ 20\% reductions}} P_0 = (0.8)^n P_0.$$

We see from its formula that $f(n)$ is an exponential function, because it is of the form $f(n) = ab^n$, with $a = P_0$ and $b = 0.8$. The graph of $y = f(n) = P_0(0.8)^n$, for $n \geq 0$, is given in Figure 4.1. Beginning at the P-axis, the curve decreases sharply at first towards the horizontal axis, but then levels off so that its descent is less rapid.

Figure 4.1 shows that the prevalence of the virus in the population drops quickly at first, and that it eventually levels off and approaches zero. The curve has this shape because in the early years of the vaccination program, there was a relatively large number of infected people. In later years, due to the success of the vaccine, the infection became increasingly rare. Thus, in the early years, a 20% drop in the infected population represented a larger number of people than a 20% drop in later years.

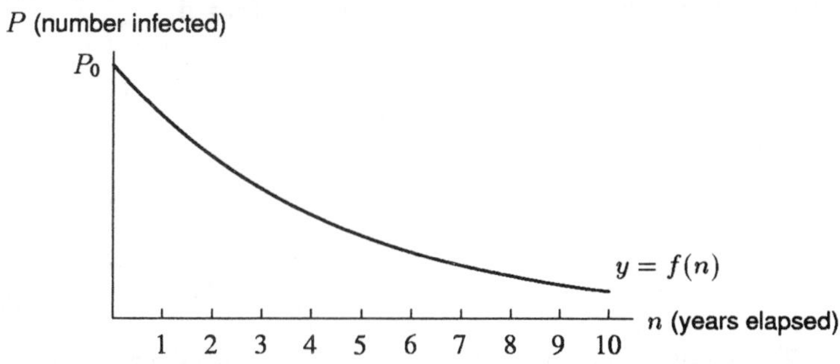

Figure 4.1: The graph of $f(n) = P_0(0.8)^n$ for $n \geq 0$

9. If an investment decreases by 5% each year, we know that only 95% remains at the end of the first year. After 2 years there will be 95% of 95%, or 0.95^2 left. After 4 years, there will be $0.95^4 \approx 0.8145$ or 81.45% of the investment left; it therefore decreases by about 18.55% altogether.

13. Using the formula for slope, we have

$$\text{Slope} = \frac{f(5)-f(1)}{5-1} = \frac{4b^5 - 4b^1}{4} = \frac{4b(b^4-1)}{4} = b(b^4-1).$$

17. (a) $N(0)$ gives the number of teams remaining in the tournament after no rounds have been played. Thus, $N(0) = 64$. After 1 round, half of the original 64 teams remain in the competition, so

$$N(1) = 64(\frac{1}{2}).$$

After 2 rounds, half of these teams remain, so

$$N(2) = 64(\frac{1}{2})(\frac{1}{2}).$$

And, after r rounds, the original pool of 64 teams has been halved r times, so that

$$N(r) = 64 \underbrace{(\frac{1}{2})(\frac{1}{2})\cdots(\frac{1}{2})}_{\text{pool halved } r \text{ times}},$$

giving

$$N(r) = 64(\frac{1}{2})^r.$$

The graph of $y = N(r)$ is given in Figure 4.2. The domain of N is $0 \le r \le 6$, for r an integer. A curve has been dashed in to help you see the overall shape of the function.

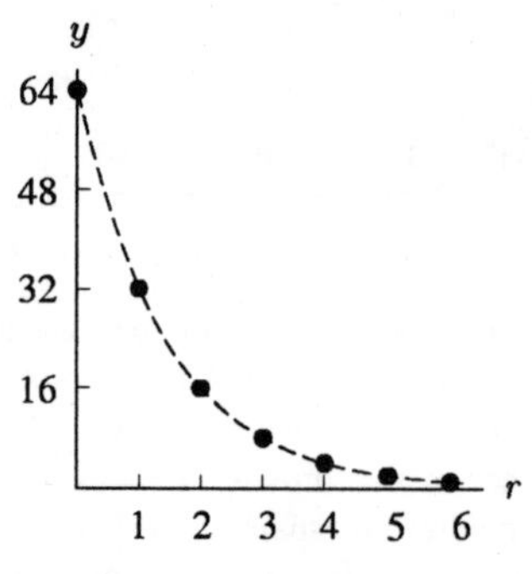

Figure 4.2: The graph of $y = N(r) = 64 \cdot \left(\frac{1}{2}\right)^r$

(b) There will be a winner when there is only one person left. So, $N(r) = 1$.

$$\begin{aligned} 64(\frac{1}{2})^r &= 1 \\ \left(\frac{1}{2}\right)^r &= \frac{1}{64} \\ \frac{1}{2^r} &= \frac{1}{64} \\ 2^r &= 64 \\ r &= 6 \end{aligned}$$

You can solve $2^r = 64$ either by taking successive powers of 2 until you get to 64 or by substituting values for r until you get the one that works.

Solutions for Section 4.2

1. If a function is linear and the x-values are equally spaced, you get from one y-value to the next by adding (or subtracting) the same amount each time. On the other hand, if the function is exponential and the x-values are evenly spaced, you get from one y-value to the next by multiplying by the same factor each time.

5. Since $f(x) = ab^x$, $f(3) = ab^3$ and $f(-2) = ab^{-2}$. Since we know that $f(3) = -\frac{3}{8}$ and $f(-2) = -12$, we can say

$$ab^3 = -\frac{3}{8}$$

and

$$ab^{-2} = -12.$$

Forming ratios, we have

$$\begin{aligned} \frac{ab^3}{ab^{-2}} &= \frac{-\frac{3}{8}}{-12} \\ b^5 &= -\frac{3}{8} \times -\frac{1}{12} = \frac{1}{32}. \end{aligned}$$

Since $32 = 2^5$, $\frac{1}{32} = \frac{1}{2^5} = (\frac{1}{2})^5$. This tells us that

$$b = \frac{1}{2}.$$

Thus, our formula is $f(x) = a(\frac{1}{2})^x$. Use $f(3) = a(\frac{1}{2})^3$ and $f(3) = -\frac{3}{8}$ to get

$$\begin{aligned} a(\frac{1}{2})^3 &= -\frac{3}{8} \\ a(\frac{1}{8}) &= -\frac{3}{8} \\ \frac{a}{8} &= -\frac{3}{8} \\ a &= -3. \end{aligned}$$

Therefore $f(x) = -3(\frac{1}{2})^x$.

9. If the function is exponential, its formula is of the form $y = ab^x$. Since $(0, 1)$ is on the graph

$$\begin{aligned} y &= ab^x \\ 1 &= ab^0 \end{aligned}$$

Since $b^0 = 1$,

$$\begin{aligned} 1 &= a(1) \\ a &= 1. \end{aligned}$$

Since $(2, 100)$ is on the graph and $a = 1$,

$$\begin{aligned} y &= ab^x \\ 100 &= (1)b^2 \\ b^2 &= 100 \\ b = 10 &\text{ or } b = -10 \end{aligned}$$

$b = -10$ is excluded, since b must be greater than zero. Therefore, $y = 1(10)^x$ or $y = 10^x$ is a possible formula for this function.

13. Since the function is exponential, we know that $y = ab^x$. The points $(-2, 45/4)$ and $(1, 10/3)$ are on the graph so,

$$\frac{45}{4} = ab^{-2}$$
$$\frac{10}{3} = ab^{1}$$

Taking the ratio of the second equation to the first one we have

$$\frac{10/3}{45/4} = \frac{ab^1}{ab^{-2}}.$$

Since $\frac{10}{3}/\frac{45}{4} = \frac{10}{3} \cdot \frac{4}{45} = \frac{8}{27}$,

$$\frac{8}{27} = b^3.$$

Since $8 = 2^3$ and $27 = 3^3$, we know that $\frac{8}{27} = \frac{2^3}{3^3} = (\frac{2}{3})^3$, so

$$(\frac{2}{3})^3 = b^3$$
$$b = \frac{2}{3}.$$

Substituting this value of b into the second equation gives

$$\frac{10}{3} = a(\frac{2}{3})^1$$
$$\frac{2}{3}a = \frac{10}{3}$$
$$a = 5.$$

Thus, $y = 5\left(\frac{2}{3}\right)^x$.

17. (a) If a function is linear, then the differences in successive function values will be constant. If a function is exponential, the ratios of successive function values will remain constant. Now

$$g(1) - g(0) = 2 - 0 = 2$$

and

$$g(2) - g(1) = 4 - 2 = 2.$$

Checking the rest of the data, we see that the differences remain constant, so $g(x)$ is linear.

(b) We know that $g(x)$ is linear, so it must be of the form

$$g(x) = b + mx$$

where m is the slope and b is the y-intercept. Since at $x = 0$, $g(0) = 0$, we know that the y-intercept is 0, so $b = 0$. Using the points $(0, 0)$ and $(1, 2)$, the slope is

$$m = \frac{2-0}{1-0} = 2.$$

Thus,

$$g(x) = 0 + 2x = 2x.$$

The graph of $y = g(x)$ is shown in Figure 4.3.

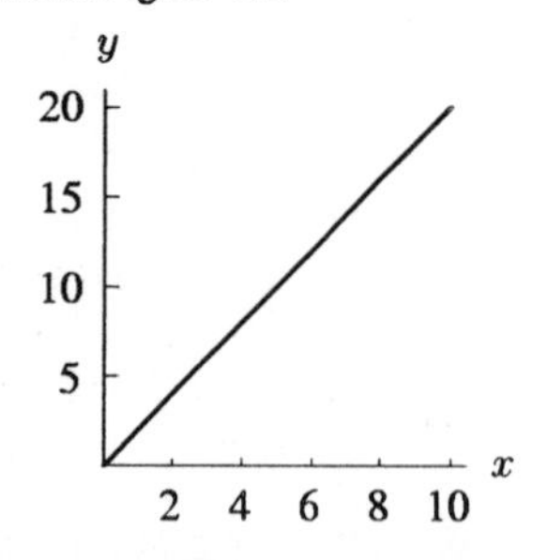

Figure 4.3

21. To use the ratio method we must have the y-values given at equally spaced x-values, which they are not. However, some of them are spaced 1 apart, namely, 1 and 2; 4 and 5; and 8 and 9. Thus, we can use these values, and consider

$$\frac{f(2)}{f(1)}, \frac{f(5)}{f(4)}, \text{ and } \frac{f(9)}{f(8)}.$$

We find

$$\frac{f(2)}{f(1)} = \frac{f(5)}{f(4)} = \frac{f(9)}{f(8)} = \frac{1}{4}.$$

With $f(x) = ab^x$ we also have

$$\frac{f(2)}{f(1)} = \frac{f(5)}{f(4)} = \frac{f(9)}{f(8)} = b,$$

so $b = \frac{1}{4}$. Using $f(1) = 4096$ we find $4096 = ab = a\left(\frac{1}{4}\right)$, so $a = 16{,}384$. Thus, $f(x) = 16{,}384\left(\frac{1}{4}\right)^x$.

25. (a) $f(0) = 1000(1.04)^0 = 1000$, which means there are 1000 people in year 0.
$f(10) = 1000(1.04)^{10} \approx 1480$, which means there are 1480 people in year 10.

(b) For the first 10 years, use $0 \le t \le 10$, $0 \le P \le 1500$. See Figure 4.4. For the first 50 years, use $0 \le t \le 50$, $0 \le P \le 8000$. See Figure 4.5.

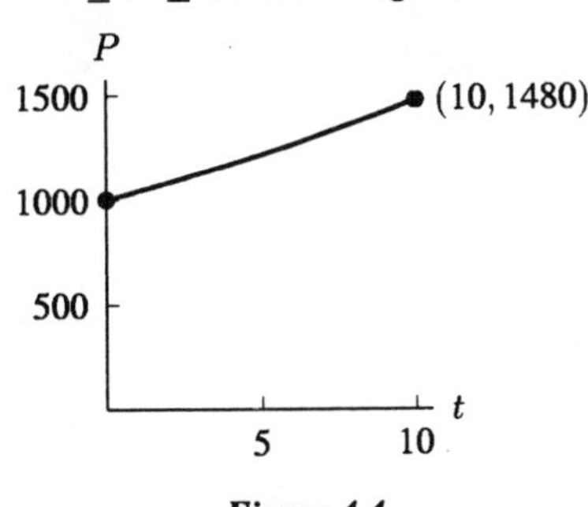

Figure 4.4

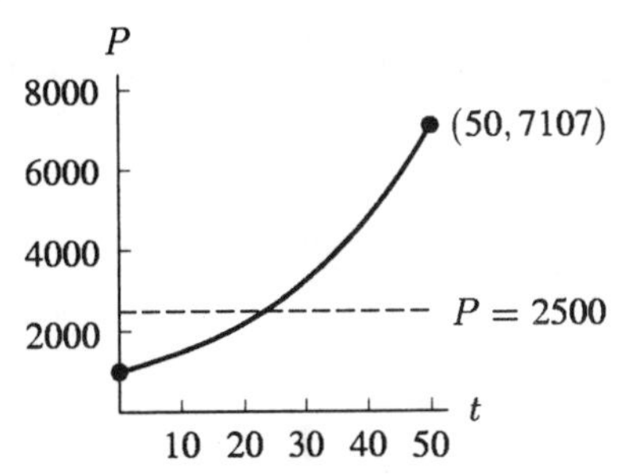

Figure 4.5

(c) The graph of $P = f(t)$ and $P = 2500$ intersect at $t \approx 23.4$. Thus, about 23.4 years after $t = 0$, the population will be 2500.

29. (a) Take 1999 to be the time $t = 0$ where t is measured in years. If V is the value, and V and t are related exponentially, then

$$V = ab^t.$$

If $V = 39,375$ at time $t = 0$, then $a = 39,375$, so

$$V = (39375)b^t.$$

We find b by calculating another point that would be on the graph of V. If the car depreciates 46% during its first 7 years, then its value when $t = 7$ is 54% of the initial price. This is $(0.54)(\$39375) = \$21262.50$. So we have the data point $(7, 21262.5)$. To find b:

$$\begin{aligned} 21262.5 &= (39375)b^7 \\ 0.54 &= b^7 \\ b &= (0.54)^{1/7} \approx 0.9157. \end{aligned}$$

So the exponential formula relating price and time is:

$$V = (39375)(0.9157)^t.$$

(b) If the depreciation is linear, then the value of the car at time t is

$$V = b + mt$$

where b is the value at time $t = 0$ (the year 1999). So $b = 39375$. We already calculated the value of the car after 7 years to be $(0.54)(\$39375) = \$21262.50$. Since $V = 21262.5$ when $t = 7$, and $b = 39375$, we have

$$\begin{aligned} 21262.50 &= 39375 + 7m, \\ -18112.5 &= 7m \\ -2587.5 &= m. \end{aligned}$$

So $V = 39,375 - 2587.5t$.

(c) Using the exponential model, the value of the car after 4 years would be:

$$V = (39375)(0.9157)^4 \approx \$27,684.29.$$

Using the linear model, the value would be:

$$V = 39{,}375 - (2587.5)(4) = \$29{,}025.$$

So the linear model would result in a higher resale price and would therefore be preferable.

Solutions for Section 4.3

1. (a)

TABLE 4.1

x	-3	-2	-1	0	1	2	3
$f(x)$	1/8	1/4	1/2	1	2	4	8

(b) For large negative values of x, $f(x)$ is close to the x-axis. But for large positive values of x, $f(x)$ climbs rapidly away from the x-axis. As x gets larger, y grows more and more rapidly. See Figure 4.6.

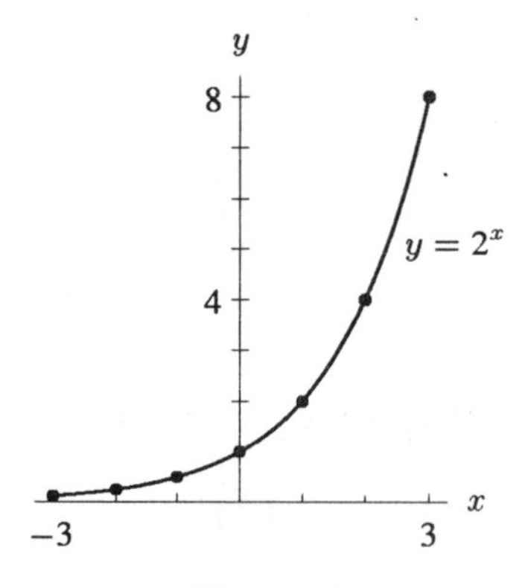

Figure 4.6

5. Let $f(x) = (0.7)^x$, $g(x) = (0.8)^x$, and $h(x) = (0.85)^x$. We note that for $x = 0$,

$$f(x) = g(x) = h(x) = 1.$$

On the other hand, $f(1) = 0.7$, $g(1) = 0.8$, and $h(1) = 0.85$, while $f(2) = 0.49$, $g(2) = 0.64$, and $h(2) = 0.7225$; so

$$0 < f(x) < g(x) < h(x).$$

So the graph of $f(x)$ lies below the graph of $g(x)$, which in turn lies below the graph of $h(x)$.

Alternately, you can consider 0.7, 0.8, and 0.85 as growth factors (decaying). The $f(x) = (0.7)^x$ will be the lowest graph because it is decaying the fastest. The $h(x) = (0.85)^x$ will be the top graph because it decays the least.

9. (a) Concave up.
(b) Concave down.
(c) Concave down.
(d) Neither.

13. Answers will vary, but they should mention that $f(x)$ is increasing and $g(x)$ is decreasing, that they have the same domain, range, and horizontal asymptote. Some may see that $g(x)$ is a reflection of $f(x)$ about the y-axis whenever $b = 1/a$. Graphs might resemble the following:

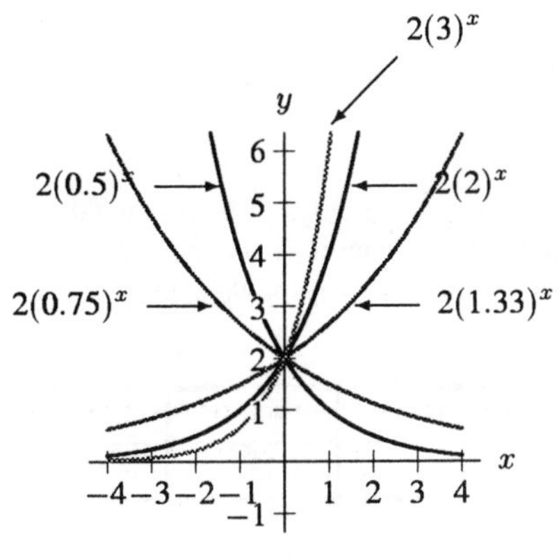

Figure 4.7

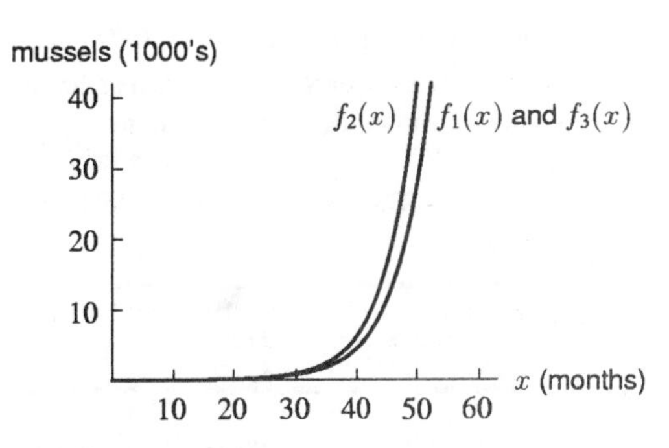

Figure 4.8

17. (a) Figure 4.8 shows the three populations. From this graph, the three models seem to be in good agreement. Models 1 and 3 are indistinguishable; model 2 appears to rise a little faster. However notice that we cannot see the behavior beyond 50 months because our function values go beyond the top of the viewing window.

(b) Figure 4.9 shows the population differences. The graph of $y = f_2(x) - f_1(x) = 3(1.21)^x - 3(1.2)^x$ grows very rapidly, especially after 40 months. The graph of $y = f_3(x) - f_1(x) = 3.01(1.2)^x - 3(1.2)^x$ is hardly visible on this scale.

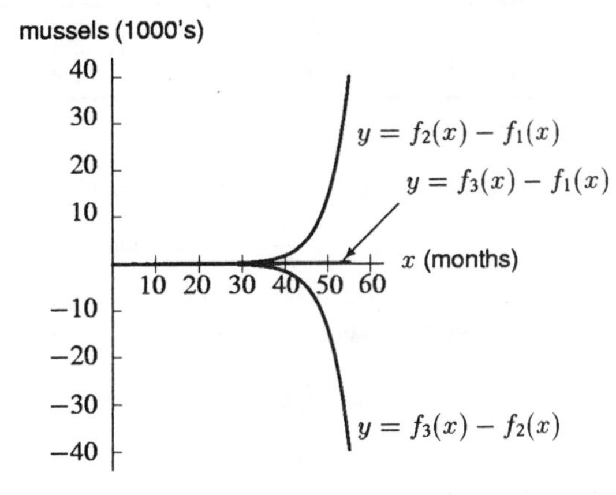

Figure 4.9

(c) Models 1 and 3 are in good agreement, but model 2 predicts a much larger mussel population than does model 1 after only 50 months. We can come to at least two conclusions. First, even small differences in the base of an exponential function can be highly significant, while differences in initial values are not as significant. Second, although two exponential curves can look very similar, they can actually be making very different predictions as time increases.

21. Since the function is exponential, we know $y = ab^x$. The points $(0, 4)$ and $(2, 4e^2)$ are on the graph, so $4 = ab^0$. Since $b^0 = 1$, we know that

$$a = 4.$$

Using the point $(2, 4e^2)$, we get:

$$4e^2 = 4b^2$$
$$e^2 = b^2.$$

Since $b > 0$, we have

$$b = e.$$

Therefore, $y = 4e^x$ is a possible formula for this function.

25. (a) At the end of 100 years,

$$B = 1200e^{0.03(100)} = 24{,}102.64 \text{ dollars}.$$

(b) Tracing along a graph of $B = 1200e^{0.03t}$ until $B = 50000$ gives $t \approx 124$ years.

29. (a) This function appears to be decreasing throughout.

(b) Table 4.2 shows the change in the number of military personnel on active duty, based on the table given in the question. Notice that the rate of change gets more negative until 1992, when it slows down. After 1992, the rate ofchange gets less negative, i.e. more positive. Thus, the graph is concave down and then concave up.

TABLE 4.2

Year	1990-91	1991-92	1992-93	1993-94	1994-95	1995-96
Change (in thousands)	−58	−179	−102	−94	−93	−46

(c) The graph changes concavity somewhere around 1992.

Solutions for Section 4.4

1. See Table 4.3. From the table, we see that the value of 10^n draws quite close to 3000 as n draws close to 3.47712, and so we estimate that $\log 3000 \approx 3.47712$. Using a calculator, we see that $\log 3000 = 3.4771212\ldots$.

TABLE 4.3

n	3	3.5	3.48	3.477	3.4771	3.47712
10^n	1000	3162.28	3019.95	2999.16	2999.85	2999.99

5. (a) Patterns:

$$\begin{aligned} \log(A \cdot B) &= \log A + \log B \\ \log \frac{A}{B} &= \log A - \log B \\ \log A^B &= B \log A \end{aligned}$$

(b) Using our patterns, we can rewrite the expression as follows,

$$\log\left(\frac{AB}{C}\right)^p = p\log\left(\frac{AB}{C}\right) = p(\log(AB) - \log C) = p(\log A + \log B - \log C).$$

9. (a)

$$\begin{aligned} \log 100^x &= \log(10^2)^x \\ &= \log 10^{2x}. \end{aligned}$$

Since $\log 10^N = N$, then

$$\log 10^{2x} = 2x.$$

(b)

$$\begin{aligned} 1000^{\log x} &= (10^3)^{\log x} \\ &= (10^{\log x})^3 \end{aligned}$$

Since $10^{\log x} = x$ we know that

$$(10^{\log x})^3 = (x)^3 = x^3.$$

(c)

$$\begin{aligned} \log 0.001^x &= \log\left(\frac{1}{1000}\right)^x \\ &= \log(10^{-3})^x \\ &= \log 10^{-3x} \\ &= -3x. \end{aligned}$$

13. Using the log rules, we have

$$\begin{aligned}
4(1.171)^x &= 7(1.088)^x \\
\frac{(1.171)^x}{(1.088)^x} &= \frac{7}{4} \\
\left(\frac{1.171}{1.088}\right)^x &= \frac{7}{4} \\
\log\left(\frac{1.171}{1.088}\right)^x &= \log\left(\frac{7}{4}\right) \\
x\log\left(\frac{1.171}{1.088}\right) &= \log\left(\frac{7}{4}\right) \\
x &= \frac{\log(7/4)}{\log(1.171/1.088)}.
\end{aligned}$$

Checking the answer with a calculator, we get

$$x = \frac{\log(7/4)}{\log(1.171/1.088)} \approx 7.6,$$

and we see that

$$4(1.171)^{7.6} \approx 13.3 \qquad 7(1.088)^{7.6} \approx 13.3.$$

17. Taking natural logs, we get

$$\begin{aligned}
e^{x+5} &= 7\cdot 2^x \\
\ln e^{x+5} &= \ln(7\cdot 2^x) \\
x+5 &= \ln 7 + \ln 2^x \\
x+5 &= \ln 7 + x\ln 2 \\
x - x\ln 2 &= \ln 7 - 5 \\
x(1-\ln 2) &= \ln 7 - 5 \\
x &= \frac{\ln 7 - 5}{1-\ln 2}
\end{aligned}$$

21. Taking logs and using the log rules:

$$\begin{aligned}
\log(ab^x) &= \log c \\
\log a + \log b^x &= \log c \\
\log a + x\log b &= \log c \\
x\log b &= \log c - \log a \\
x &= \frac{\log c - \log a}{\log b}.
\end{aligned}$$

25. Since the goal is to get t by itself as much as possible, first divide both sides by 3, and then use logs.

$$\begin{aligned}
3(1.081)^t &= 14 \\
1.081^t &= \frac{14}{3} \\
\log(1.081)^t &= \log(\frac{14}{3}) \\
t\log 1.081 &= \log(\frac{14}{3}) \\
t &= \frac{\log(\frac{14}{3})}{\log 1.081} \approx 19.8
\end{aligned}$$

29.
$$
\begin{aligned}
121e^{-0.112t} &= 88 \\
e^{-0.112t} &= \frac{88}{121} \\
\ln e^{-0.112t} &= \ln(\frac{88}{121}) \\
-0.112t &= \ln(\frac{88}{121}) \\
t &= \frac{\ln(88/121)}{-0.112} \approx 2.84
\end{aligned}
$$

33. $10^{1.3} \approx 20$ tells us that $\log 20 \approx 1.3$. Using one of the properties of logarithms, we can find $\log 200$:
$$\log 200 = \log(10 \cdot 20) = \log 10 + \log 20 \approx 1 + 1.3 = 2.3.$$

Solutions for Section 4.5

1. The graphs of $y = 10^x$ and $y = 2^x$ both have horizontal asymptotes, $y = 0$. The graph of $y = \log x$ has a vertical asymptote, $x = 0$.

5. The graphs is:

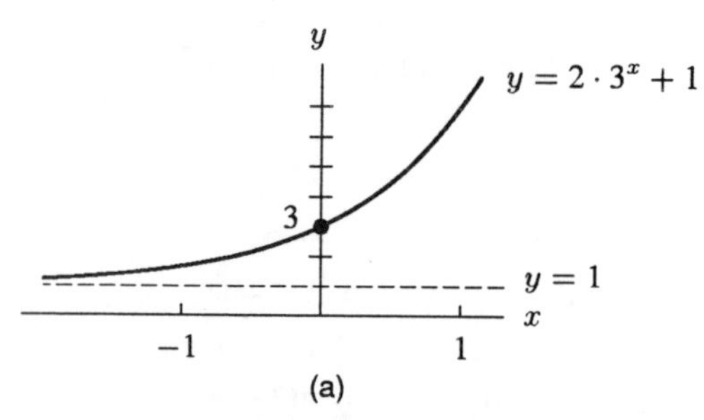

(a)

The graph of $y = 2 \cdot 3^x + 1$ is the graph of $y = 3^x$ stretched vertically by a factor of 2 and shifted up by 1 unit.

9. (a) Let the functions graphed in (a), (b), and (c) be called $f(x)$, $g(x)$, and $h(x)$ respectively. Looking at the graph of $f(x)$, we see that $f(10) = 3$. In the table for $r(x)$ we note that $r(10) = 1.6990$ so $f(x) \neq r(x)$. Similarly, $s(10) = 0.6990$, so $f(x) \neq s(x)$. The values describing $t(x)$ do seem to satisfy the graph of $f(x)$, however. In the graph, we note that when $0 < x < 1$, then y must be negative. The data point $(0.1, -3)$ satisfies this. When $1 < x < 10$, then $0 < y < 3$. In the table for $t(x)$, we see that the point $(2, 0.9031)$ satisfies this condition. Finally, when $x > 10$ we see that $y > 3$. The values $(100, 6)$ satisfy this. Therefore, $f(x)$ and $t(x)$ could represent the same function.

(b) For $g(x)$, we note that
$$\begin{cases} \text{when } 0 < x < 0.2, & \text{then } y < 0; \\ \text{when } 0.2 < x < 1, & \text{then } 0 < y < 0.699; \\ \text{when } x > 1, & \text{then } y > 0.699. \end{cases}$$
All the values of x in the table for $r(x)$ are greater than 1 and all the corresponding values of y are greater than 0.699, so $g(x)$ could equal $r(x)$. We see that, in $s(x)$, the values $(0.5, -0.06021)$ do not satisfy the second condition so $g(x) \neq s(x)$. Since we already know that $t(x)$ corresponds to $f(x)$, we conclude that $g(x)$ and $r(x)$ correspond.

(c) By elimination, $h(x)$ must correspond to $s(x)$. We see that in $h(x)$,
$$\begin{cases} \text{when } x < 2, & \text{then } y < 0; \\ \text{when } 2 < x < 20, & \text{then } 0 < y < 1; \\ \text{when } x > 20, & \text{then } y > 1. \end{cases}$$
Since the values in $s(x)$ satisfy these conditions, it is reasonable to say that $h(x)$ and $s(x)$ correspond.

13. We need $x^2 > 0$, which is true as long as $x \neq 0$, so the domain is all $x \neq 0$.

17. A possible formula is $y = \log x$.

21. This graph could represent exponential decay, with a y-intercept of 0.1. A possible formula is $y = 0.1b^x$ with $0 < b < 1$.

Solutions for Section 4.6

1. Get all expressions containing x on one side of the equation and everything else on the other side. To do this we divide both sides of the equation by 1.7 and by $(4.5)^x$.

$$1.7(2.1)^{3x} = 2(4.5)^x$$
$$\frac{(2.1)^{3x}}{(4.5)^x} = \frac{2}{1.7}$$

We know that $(2.1)^{3x} = [(2.1)^3]^x = (9.261)^x$. Therefore,

$$\frac{(9.261)^x}{(4.5)^x} = \frac{2}{1.7}$$
$$\left(\frac{9.261}{4.5}\right)^x = \frac{2}{1.7}$$
$$\log\left(\frac{9.261}{4.5}\right)^x = \log\left(\frac{2}{1.7}\right)$$
$$x\log\left(\frac{9.261}{4.5}\right) = \log\left(\frac{2}{1.7}\right)$$
$$x = \frac{\log(2/1.7)}{\log(9.261/4.5)} \approx 0.225178.$$

5. First rewrite $10e^{3t} - e = 2e^{3t}$ as

$$8e^{3t} = e.$$

Then take natural logs and use the log rules

$$\ln(8e^{3t}) = \ln e$$
$$\ln 8 + \ln e^{3t} = 1$$
$$\ln 8 + 3t = 1$$
$$3t = 1 - \ln 8$$
$$t = \frac{1 - \ln 8}{3} = -0.360.$$

9. Let $P = ab^t$ where P is the number of bacteria at time t hours since the beginning of the experiment. a is the number of bacteria we're starting with.

(a) Since the colony begins with 3 bacteria we have $a = 3$. Using the information that $P = 100$ when $t = 3$, we can solve the following equation for b:

$$P = 3b^t$$
$$100 = 3b^3$$
$$\sqrt[3]{\frac{100}{3}} = b$$
$$b = \left(\frac{100}{3}\right)^{1/3} \approx 3.22$$

Therefore, $P = 3(3.22)^t$.

(b) We want to find the value of t for which the population triples, going from three bacteria to nine. So we want to solve:

$$9 = 3(3.22)^t$$
$$3 = (3.22)^t$$
$$\log 3 = \log(3.22)^t$$
$$= t\log(3.22).$$

Thus,

$$t = \frac{\log 3}{\log(3.22)} \approx 0.939 \text{ hours.}$$

13. (a) Initially, the population is $P = 300 \cdot 2^{0/20} = 300 \cdot 2^0 = 300$. After 20 years, the population reaches $P = 300 \cdot 2^{20/20} = 300 \cdot 2^1 = 600$.

(b) To find when the population reaches $P = 1000$, we solve the equation:

$$\begin{aligned} 300 \cdot 2^{t/20} &= 1000 \\ 2^{t/20} &= \frac{1000}{300} = \frac{10}{3} && \text{dividing by 300} \\ \log\left(2^{t/20}\right) &= \log\left(\frac{10}{3}\right) && \text{taking logs} \\ \left(\frac{t}{20}\right) \cdot \log 2 &= \log\left(\frac{10}{3}\right) && \text{using a log property} \\ t &= \frac{20\log(10/3)}{\log 2} = 34.7, \end{aligned}$$

and so it will take the population a bit less than 35 years to reach 1000.

17. Let t be the doubling time, then the population is $2P_0$ at time t, so

$$\begin{aligned} 2P_0 &= P_0 e^{0.2t} \\ 2 &= e^{0.2t} \\ 0.2t &= \ln 2 \\ t &= \frac{\ln 2}{0.2} \approx 3.466. \end{aligned}$$

21. (a) Since $f(x)$ is exponential, its formula will be $f(x) = ab^x$. Since $f(0) = 0.5$,

$$f(0) = ab^0 = 0.5.$$

But $b^0 = 1$, so

$$\begin{aligned} a(1) &= 0.5 \\ a &= 0.5. \end{aligned}$$

We now know that $f(x) = 0.5b^x$. Since $f(1) = 2$, we have

$$\begin{aligned} f(1) = 0.5b^1 &= 2 \\ 0.5b &= 2 \\ b &= 4 \end{aligned}$$

So $f(x) = 0.5(4)^x$.

We will find a formula for $g(x)$ the same way.

$$g(x) = ab^x.$$

Since $g(0) = 4$,

$$\begin{aligned} g(0) = ab^0 &= 4 \\ a &= 4. \end{aligned}$$

Therefore,

$$g(x) = 4b^x.$$

We'll use $g(2) = \frac{4}{9}$ to get

$$\begin{aligned} g(2) = 4b^2 &= \frac{4}{9} \\ b^2 &= \frac{1}{9} \\ b &= \pm\frac{1}{3}. \end{aligned}$$

Since $b > 0$,

$$g(x) = 4\left(\frac{1}{3}\right)^x.$$

Since $h(x)$ is linear, its formula will be

$$h(x) = b + mx.$$

We know that b is the y-intercept, which is 2, according to the graph. Since the points $(a, a+2)$ and $(0, 2)$ lie on the graph, we know that the slope, m, is

$$\frac{(a+2)-2}{a-0} = \frac{a}{a} = 1,$$

so the formula is

$$h(x) = 2 + x.$$

(b) We begin with

$$\begin{aligned} f(x) &= g(x) \\ \frac{1}{2}(4)^x &= 4\left(\frac{1}{3}\right)^x. \end{aligned}$$

Since the variable is an exponent, we need to use logs, so

$$\begin{aligned} \log\left(\frac{1}{2}\cdot 4^x\right) &= \log\left(4\cdot\left(\frac{1}{3}\right)^x\right) \\ \log\frac{1}{2} + \log(4)^x &= \log 4 + \log\left(\frac{1}{3}\right)^x \\ \log\frac{1}{2} + x\log 4 &= \log 4 + x\log\frac{1}{3}. \end{aligned}$$

Now we will move all expressions containing the variable to one side of the equation:

$$x\log 4 - x\log\frac{1}{3} = \log 4 - \log\frac{1}{2}.$$

Factoring out x, we get

$$\begin{aligned} x(\log 4 - \log\frac{1}{3}) &= \log 4 - \log\frac{1}{2} \\ x\log\left(\frac{4}{1/3}\right) &= \log\left(\frac{4}{1/2}\right) \\ x\log 12 &= \log 8 \\ x &= \frac{\log 8}{\log 12}. \end{aligned}$$

This is the exact value of x. Note that $\frac{\log 8}{\log 12} \approx 0.837$, so $f(x) = g(x)$ when x is exactly $\frac{\log 8}{\log 12}$ or about 0.837.

(c) Since $f(x) = h(x)$, we want to solve

$$\frac{1}{2}(4)^x = x + 2.$$

The variable does not occur only as an exponent, so logs cannot help us solve this equation. Instead, we need to graph the two functions and note where they intersect. The points occur when $x \approx 1.38$ or $x \approx -1.97$.

25. (a) We know that $D_1 = 10\log\left(\frac{I_1}{I_0}\right)$ and $D_2 = 10\log\left(\frac{I_2}{I_0}\right)$. Thus

$$\begin{aligned} D_2 - D_1 &= 10\log\left(\frac{I_2}{I_0}\right) - 10\log\left(\frac{I_1}{I_0}\right) \\ &= 10\left(\log\left(\frac{I_2}{I_0}\right) - \log\left(\frac{I_1}{I_0}\right)\right) && \text{factoring} \\ &= 10\log\left(\frac{I_2/I_0}{I_1/I_0}\right) && \text{using a log property} \end{aligned}$$

and so

$$D_2 - D_1 = 10\log\left(\frac{I_2}{I_1}\right).$$

(b) Suppose the sound's initial intensity is I_1 and that its new intensity is I_2. Then here we have $I_2 = 2I_1$. If D_1 is the original decibel rating and D_2 is the new rating then

$$\begin{aligned} \text{Increase in decibels} &= D_2 - D_1 \\ &= 10\log\left(\frac{I_2}{I_1}\right) && \text{using formula from part (a)} \\ &= 10\log\left(\frac{2I_1}{I_1}\right) \\ &= 10\log 2 \\ &\approx 3.01. \end{aligned}$$

Thus, the sound increases by 3 decibels when it doubles in intensity.

Solutions for Section 4.7

1. Substituting $t = 0$ we get

$$f(0) = 5(\frac{1}{2})^0 = 5 = A_0e^0 = A_0.$$

So,

$$A_0 = 5.$$

Substituting $t = 1$ we get

$$f(1) = 5(\frac{1}{2}) = 2.5 = 5e^{-k}.$$

Dividing both sides by 5 and taking logs we get

$$\ln\frac{1}{2} = -k$$

or

$$-\ln\frac{1}{2} = k$$

or

$$k = \ln 2.$$

5. (a) Using $P = P_0e^{kt}$ where $P_0 = 25{,}000$ and $k = 7.5\%$, we have

$$P(t) = 25{,}000e^{0.075t}.$$

(b) We first need to find the growth factor so will rewrite

$$P = 25{,}000e^{0.075t} = 25{,}000(e^{0.075})^t \approx 25{,}000(1.0779)^t.$$

At the end of a year, the population is 107.79% of what it had been at the end of the previous year. This corresponds to an increase of approximately 7.79%. This is greater than 7.5% because the rate of 7.5% per year is being applied to larger and larger amounts. In one instant, the population is growing at a rate of 7.5% per year. In the next instant, it grows again at a rate of 7.5% a year, but 7.5% of a slightly larger number. The fact that the population is increasing in tiny increments continuously results in an actual increase greater than the 7.5% increase that would result from one, single jump of 7.5% at the end of the year.

9. (a) We know that $P(t) = ab^t$, and

$$P(0) = ab^0 = a.$$

We are told that $P(0) = 30$, so $a = 30$, and $P(t) = 30b^t$. But $P(10) = 45$ and $P(t) = 30b^t$, so

$$\begin{aligned} P(10) &= 30b^{10} \\ 45 &= 30b^{10} \\ \frac{45}{30} &= b^{10} \\ b &= \left(\frac{45}{30}\right)^{1/10} \approx 1.0414. \end{aligned}$$

b is the annual growth factor of the population. The population is 104.14% of what it had been the previous year, so we know that it is growing by approximately 4.14% each year.

(b) Since $P(t) = ab^t$ and $P(t) = P_0e^{kt}$, both represent the same population, and thus we know that

$$ab^t = P_0e^{kt}.$$

We know that a and P_0 both represent the initial population, so $a = P_0$. Thus

$$\begin{aligned} ab^t &= ae^{kt} \\ b^t &= e^{kt}. \end{aligned}$$

This tells us that $b = e^k$. We know that $b = (45/30)^{1/10} \approx 1.0414$, so $e^k \approx 1.0414$.

To find the value of k, we can either use trial and error or we can graph $y = e^x$ and $y = (45/30)^{1/10} \approx 1.0414$ to find the point of intersection. Either way, we learn that $k \approx 0.0405$. This value, 4.05%, is the continuous annual growth rate of the population. In other words, at any given instant, the population is growing at the rate of 4.05% per year. We note that this rate is slightly less than the actual percent increase for the year, which is 4.14%.

13. (a) If $Q(t) = Q_0 b^t$ describes the number of gallons left in the tank after t hours, then Q_0, the amount we started with, is 250, and b, the percent left in the tank after 1 hour, is 96%. Thus $Q(t) = 250(0.96)^t$. After 10 hours, there are $Q(10) = 250(0.96)^{10} \approx 166.2$ gallons left in the tank. This $\frac{166.2}{250} = 0.665 = 66.5\%$ of what had initially been in the tank. Therefore approximately 33.5% has leaked out. It is less than 40% because the loss is 4% of 250 only during the first hour; for each hour after that it is 4% of whatever quantity is left.

(b) Since $Q_0 = 250$, $Q(t) = 250e^{kt}$. But we can also define $Q(t) = 250(0.96)^t$, so

$$\begin{aligned} 250e^{kt} &= 250(0.96)^t \\ e^{kt} &= 0.96^t \\ e^k &= 0.96 \\ \ln e^k &= \ln 0.96 \\ k \ln e &= \ln 0.96 \\ k &= \ln 0.96 \approx -0.041. \end{aligned}$$

Since k is negative, we know that the value of $Q(t)$ is decreasing by 4.1% per hour. Therefore, k is the continuous hourly decay rate.

17. For what value of t will $Q(t) = 0.23Q_0$?

$$\begin{aligned} 0.23Q_0 &= Q_0 e^{-0.000121t} \\ 0.23 &= e^{-0.000121t} \\ \ln 0.23 &= \ln e^{-0.000121t} \\ \ln 0.23 &= -0.000121t \\ t &= \frac{\ln 0.23}{-0.000121} \approx 12146. \end{aligned}$$

So the skull is about 12,146 years old.

21. The new population is given by the formula $P_2 = 5.2(1.031)^t$. The population in the text is given by $P_1 = 7.3e^{0.022t}$. Thus, we must solve the equation

$$5.2(1.031)^t = 7.3e^{0.022t}.$$

Taking the natural log of both sides gives

$$\ln(5.2(1.031)^t) = \ln(7.3e^{0.022t}).$$

Using the properties of the natural log, we get

$$\begin{aligned} \ln 5.2 + \ln 1.031^t &= \ln 7.3 + \ln e^{0.022t} && \text{(using } \ln ab = \ln a + \ln b\text{)} \\ \ln 5.2 + t \ln 1.031 &= \ln 7.3 + 0.022t && \text{(using } \ln a^b = b \cdot \ln a\text{)} \\ t \ln 1.031 - 0.022t &= \ln 7.3 - \ln 5.2 && \text{(grouping like terms)} \\ t(\ln 1.031 - 0.022) &= \ln 7.3 - \ln 5.2 && \text{(factoring } t\text{)} \\ t &= \frac{\ln 7.3 - \ln 5.2}{\ln 1.031 - 0.022} \approx 39.8. \end{aligned}$$

Thus, the populations are equal after about 39.8 years. You can use a computer or graphing calculator to check that graphs of the two functions intersect at $t \approx 39.8$.

25. (a) The probability of failure within 6 months is

$$P(6) = 1 - e^{(-0.016)(6)} \approx 0.0915 = 9.15\%.$$

In order to find the probability of failure in the second six months, we must first find the probability of its failure in the first 12 months and then subtract the probability of failure in the first six months. The probability of failure within the first 12 months is

$$P(12) = 1 - e^{(-0.016)(12)} \approx 0.1747 = 17.47\%.$$

Therefore, the probability of failure within the second 6 months is

$$17.47 - 9.15 = 8.32\%.$$

(b) We want to find t such that

$$\begin{aligned} 1-e^{-0.016t} &= 99.99\% \\ 1-e^{-0.016t} &= 0.9999 \\ e^{-0.016t} &= 0.0001 \\ \frac{1}{e^{0.016t}} &= \frac{1}{10{,}000} \\ 10{,}000 &= e^{0.016t}. \end{aligned}$$

Taking the ln of both sides and solving for t we get

$$t = \frac{\ln 10{,}000}{0.016}.$$

We see that $t \approx 576$ months, or 48 years.

Solutions for Section 4.8

1. (a) The nominal interest rate is 8%, so the interest rate per month is $0.08/12$. Therefore, at the end of 3 years, or 36 months,

$$\text{Balance} = \$1000\left(1+\frac{0.08}{12}\right)^{36} = \$1270.24.$$

(b) There are 52 weeks in a year, so the interest rate per week is $0.08/52$. At the end of $52 \times 3 = 156$ weeks,

$$\text{Balance} = \$1000\left(1+\frac{0.08}{52}\right)^{156} = \$1271.01.$$

(c) Assuming no leap years, the interest rate per day is $0.08/365$. At the end of 3×365 days

$$\text{Balance} = \$1000\left(1+\frac{0.08}{365}\right)^{3\cdot 365} = \$1271.22.$$

(d) With continuous compounding, after 3 years

$$\text{Balance} = \$1000e^{0.08(3)} = \$1271.25$$

5. Since the first student's \$500 is growing by a factor of 1.045 each year (100% + 4.5%), a formula that describes how much money she has at the end of t years is $A_1 = 500(1.045)^t$. A formula for the second student's investment is $A_2 = 800(1.03)^t$. We need to find the value of t for which $A_1 = A_2$. That is, when

$$\begin{aligned} 500(1.045)^t &= 800(1.03)^t \\ \frac{1.045^t}{1.03^t} &= \frac{800}{500} \\ \left(\frac{1.045}{1.03}\right)^t &= \frac{8}{5} \\ \log\left(\frac{1.045}{1.03}\right)^t &= \log\left(\frac{8}{5}\right) \\ t\log\left(\frac{1.045}{1.03}\right) &= \log\left(\frac{8}{5}\right) \\ t &= \frac{\log\left(\frac{8}{5}\right)}{\log\left(\frac{1.045}{1.03}\right)} \approx 32.5. \end{aligned}$$

The balances will be equal in about 32.5 years.

9. (a) If the money is compounded monthly, the interest rate is $\frac{6\%}{12}$ each month. In t years, there are $12t$ months, so the formula describing the amount of money in the account is

$$V(t) = 1000(1+\frac{0.06}{12})^{12t} = 1000(1.005)^{12t}.$$

(b) At the end of 1 year, the amount of money in the account is

$$V(1) = 1000(1.005)^{12} = 1000(1.005^{12}).$$

We want to find the continuous rate that would result in the same amount. In other words, for what value of k would the following equation hold true?

$$\begin{aligned} V(1) = 1000e^k &= 1000(1.005^{12}) \\ e^k &= 1.005^{12} \\ \ln e^k &= \ln 1.005^{12} \\ k &= 12\ln(1.005) \approx 0.0599 \end{aligned}$$

So a continuous rate of 5.99% gives the same result as 6% compounded monthly.

13. (a) Suppose \$1 is put in the account. The interest rate per month is $0.08/12$. At the end of a year,

$$\text{Balance} = \left(1 + \frac{0.08}{12}\right)^{12} = \$1.08300,$$

which is 108.3% of the original amount. So the effective annual yield is 8.300%.

(b) With weekly compounding, the interest rate per week is $0.08/52$. At the end of a year,

$$\text{Balance} = \left(1 + \frac{0.08}{52}\right)^{52} = \$1.08322,$$

which is 108.322% of the original amount. So the effective annual yield is 8.322%.

(c) Assuming it is not a leap year, the interest rate per day is $0.08/365$. At the end of a year

$$\text{Balance} = \left(1 + \frac{0.08}{365}\right)^{365} = \$1.08328,$$

which is 108.328% of the original amount. So the effective annual yield is 8.328%.

(d) For continuous compounding, at the end of the year

$$\text{Balance} = e^{0.08} = \$1.08329,$$

which is 108.329% of the original amount. So the effective annual yield is 8.329%.

17. (a) For investment A, we have

$$P = 875(1 + \frac{0.135}{365})^{365(2)} = \$1146.16.$$

For investment B,

$$P = 1000(e^{0.067(2)}) = \$1143.39.$$

For investment C,

$$P = 1050(1 + \frac{0.045}{12})^{12(2)} = \$1148.69.$$

(b) A comparison of final balances does not reflect the fact that the initial investment amounts are different. One way to take initial amount into consideration is to look at the overall growth in the account. Comparing final balance to initial deposit for each account we find

$$\text{Investment A: } \frac{1146.16}{875} \approx 1.31$$

$$\text{Investment B: } \frac{1143.39}{1000} \approx 1.143$$

$$\text{Investment C: } \frac{1148.69}{1050} \approx 1.093.$$

Thus, in the two year period Investment A has grown by approximately 31%, followed by Investment B (14.3%) and finally Investment C (9.3%). From best to worst, we have A, B, C.

[Note: Comparing the effective annual rates for each account would be a more efficient way to solve the problem and would give the same result.]

21. Let $Q = ae^{kt}$ be an increasing exponential function, so that k is positive. To find the doubling time, we find how long it takes Q to double from its initial value a to the value $2a$:

$$\begin{aligned} ae^{kt} &= 2a \\ e^{kt} &= 2 \quad \text{(dividing by } a\text{)} \\ \ln e^{kt} &= \ln 2 \\ kt &= \ln 2 \quad \text{(because } \ln e^x = x \text{ for all } x\text{)} \\ t &= \frac{\ln 2}{k}. \end{aligned}$$

Using a calculator, we find $\ln 2 = 0.693 \approx 0.70$. This is where the 70 comes from.

If, for example, the continuous growth rate is $k = 0.07 = 7\%$, then

$$\text{Doubling time} = \frac{\ln 2}{0.07} = \frac{0.693}{0.07} \approx \frac{0.70}{0.07} = \frac{70}{7} = 10.$$

If the growth rate is $r\%$, then $k = r/100$. Therefore

$$\text{Doubling time} = \frac{\ln 2}{k} = \frac{0.693}{k} \approx \frac{0.70}{r/100} = \frac{70}{r}.$$

Solutions for Section 4.9

1. The Declaration of Independence was signed in 1776, about 225 years ago. We can write this number as

$$\frac{225}{1{,}000{,}000} = 0.000225 \text{ million years ago.}$$

This number is between $10^{-4} = 0.0001$ and $10^{-3} = 0.001$. Using a calculator, we have

$$\log 0.000225 \approx -3.65,$$

which, as expected, lies between -3 and -4 on the log scale. Thus, the Declaration of Independence is placed at

$$10^{-3.65} \approx 0.000224 \text{ million years ago} = 224 \text{ years ago.}$$

5. (a) Table 4.4 is the completed table.

TABLE 4.4 *The mass of various animals in kilograms*

Animal	Body mass	log of body mass
Blue Whale	91000	4.96
African Elephant	5450	3.74
White Rhinoceros	3000	3.48
Hippopotamus	2520	3.40
Black Rhinoceros	1170	3.07
Horse	700	2.85
Lion	180	2.26
Human	70	1.85
Albatross	11	1.04
Hawk	1	0.00
Robin	0.08	−1.10
Hummingbird	0.003	−2.52

(b) Figure 4.10 shows Table 4.4 plotted on a linear scale.

Figure 4.10: On a linear scale, all masses except that of the blue whale are very close together

(c) Figure 4.11 shows Table 4.4 plotted on a logarithmic scale.

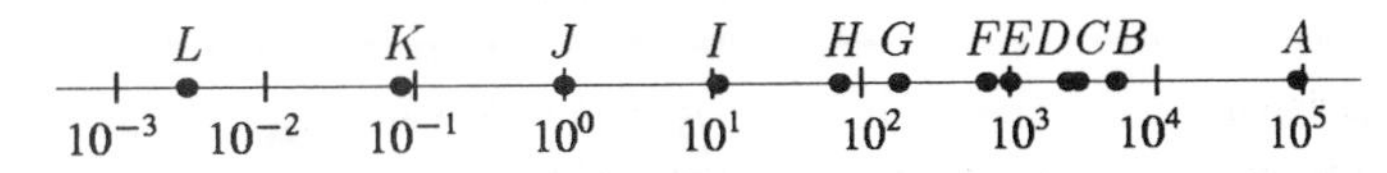

Figure 4.11

(d) Figure 4.11 gives more information than Figure 4.10.

9. Table 4.5 gives the logs of the sizes of the various organisms. The log values from Table 4.5 have been plotted in Figure 4.12.

TABLE 4.5 *Size (in cm) and* log(*size*) *of various organisms*

Animal	Size	log(size)	Animal	Size	log(size)
Virus	0.0000005	−6.3	Cat	60	1.8
Bacterium	0.0002	−3.7	Wolf	200	2.3
Human cell	0.002	−2.7	Shark	600	2.8
Ant	0.8	−0.1	Squid	2200	3.3
Hummingbird	12	1.1	Sequoia	7500	3.9

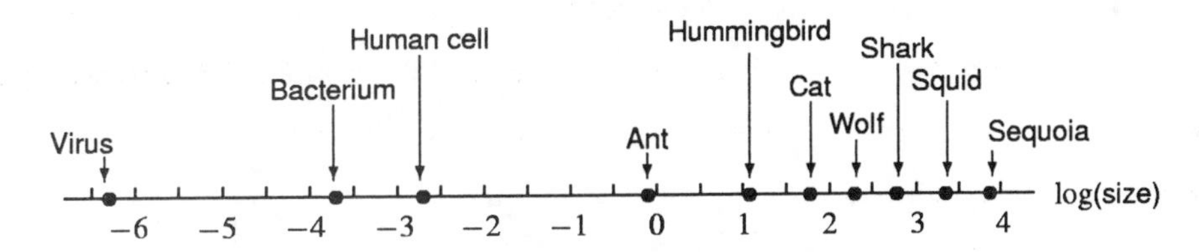

Figure 4.12: The log(sizes) of various organisms (sizes in cm)

Solutions for Section 4.10

1. (a)

TABLE 4.6

x	0	1	2	3	4	5
$y = 3^x$	1	3	9	27	81	243

(b)

TABLE 4.7

x	0	1	2	3	4	5
$y = \log(3^x)$	0	0.477	0.954	1.431	1.908	2.386

The differences between successive terms are constant(≈ 0.477), so the function is linear.

(c)

TABLE 4.8

x	0	1	2	3	4	5
$f(x)$	2	10	50	250	1250	6250

TABLE 4.9

x	0	1	2	3	4	5
$g(x)$	0.301	1	1.699	2.398	3.097	3.796

We see that $f(x)$ is an exponential function (note that it is increasing by a constant growth factor of 5), while $g(x)$ is a linear function with a constant rate of change of 0.699.

(d) The resulting function is linear. If $f(x) = a \cdot b^x$ and $g(x) = \log(a \cdot b^x)$ then

$$\begin{aligned} g(x) &= \log(ab^x) \\ &= \log a + \log b^x \\ &= \log a + x \log b \\ &= k + m \cdot x, \end{aligned}$$

where the y intercept $k = \log a$ and $m = \log b$. Thus, g will be linear.

5. (a) Run a linear regression on the data. The resulting function is $y = -3582 + 236x$, with $r \approx 0.7946$. We see from the sketch of the graph of the data that the estimated regression line provides a reasonable but not excellent fit. See Figure 4.13.

(b) If, instead, we compare x and $\ln y$ we get

$$\ln y = 1.57 + 0.2x.$$

We see from the sketch of the graph of the data that the estimated regression line provides an excellent fit with $r \approx 0.9998$. See Figure 4.14. Solving for y, we have

$$\begin{aligned} e^{\ln y} &= e^{1.57+0.2x} \\ y &= e^{1.57}e^{0.2x} \\ y &= 4.81e^{0.2x} \\ \text{or} \quad y &= 4.81(e^{0.2})^x \approx 4.81(1.22)^x. \end{aligned}$$

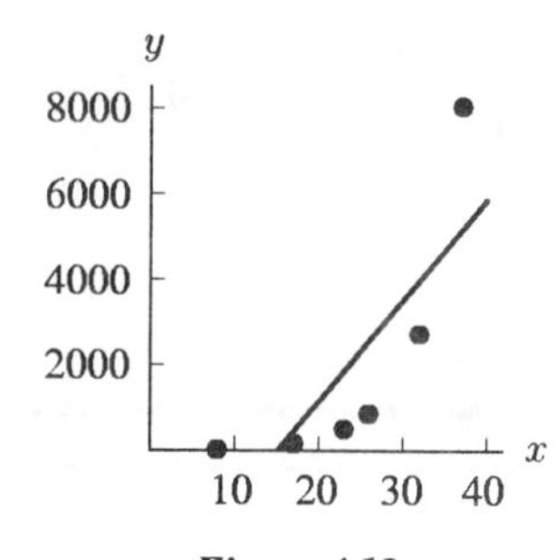

Figure 4.13

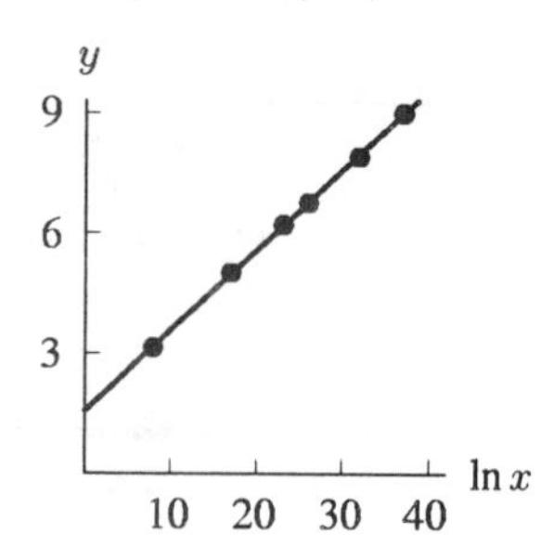

Figure 4.14

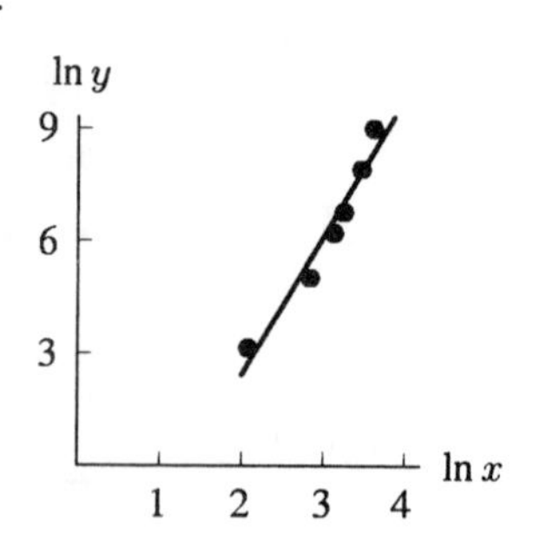

Figure 4.15

(c) Using linear regression on $\ln x$ and $\ln y$, we get

$$\ln y = -4.94 + 3.68 \ln x.$$

We see from the sketch of the graph of the data that the estimated regression line provides a good fit with $r \approx 0.9751$. See Figure 4.15. Now find the power function:

$$\begin{aligned} e^{\ln y} &= e^{-4.94+3.68 \ln x} \\ y &= e^{-4.94}e^{3.68 \ln x} \\ y &\approx 0.007(e^{\ln x})^{3.68} \\ y &\approx 0.007x^{3.68}. \end{aligned}$$

(d) The linear equation is a poor fit, and the power function is a reasonable fit, but not as good as the exponential fit.

9. (a) Because the data gives a decreasing function, we expect p to be negative.

(b) If $l = kc^p$, we want a formula using natural logs such that

$$\ln l = \ln(kc^p)$$
$$\ln l = \ln k + \ln c^p$$
$$\ln l = \ln k + p \ln c$$
$$\text{Since } y = \ln l \text{ and } x = \ln c, \quad y = \ln k + px.$$

Letting $b = \ln k$, we have $y = b + px$.

(c) We must omit the point $(0, 244)$ because $\ln 0$ is undefined. The data is in Table 4.10.

TABLE 4.10

$x = \ln c$	$y = \ln l$
−0.22	5.35
1.76	5.32
3.14	5.12
3.97	4.83
4.67	4.67
5.01	4.28
5.33	3.56
5.66	2.48
5.81	1.57
6.01	0.83
6.20	0.18

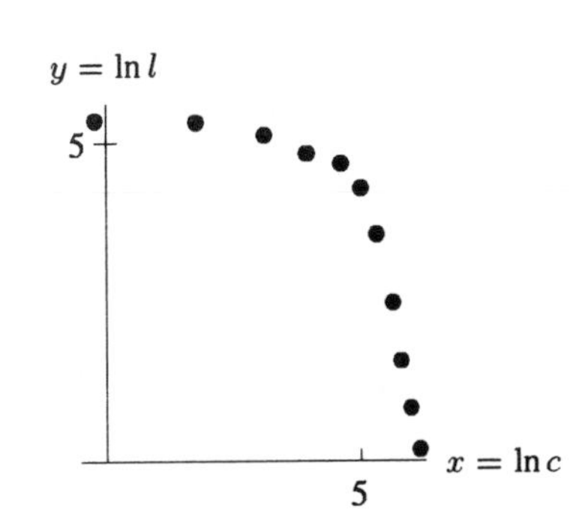

Figure 4.16

(d) The plotted points in Figure 4.16 don't look linear, so a power function won't give a good fit. From the regression equation, you could derive $k \approx 720$ and $p \approx -0.722$. If you plot $l = 720c^{-0.722}$, you will find it fits the data poorly, as predicted by the nonlinearity of the points in Figure 4.16.

13. (a) A computer or calculator gives

$$N = -14t^4 + 433t^3 - 2255t^2 + 5634t - 4397.$$

(b) The graph of the data and the quartic in Figure 4.17 shows a good fit between 1980 and 1996.

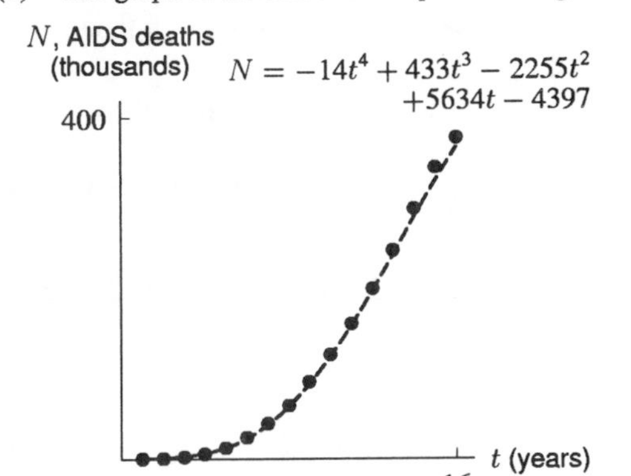

Figure 4.17

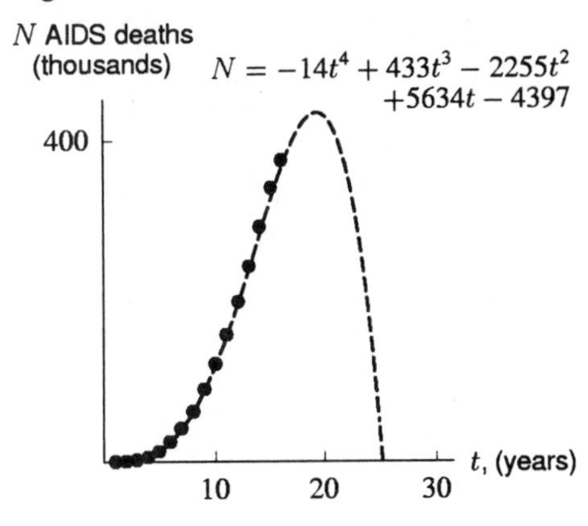

Figure 4.18

(c) Figure 4.17 shows that the quartic model fits the 1980-1996 data well. However, this model predicts that in 2000, the number of deaths decreases. See Figure 4.18. Since N is the total number of deaths since 1980, this is impossible. Therefore the quartic is definitely not a good model for $t > 20$.

Solutions for Chapter 4 Review

1. (a) After 5 years, $B = 5000(1.12)^5 = \$8{,}811.71$. After 10 years, $B = 5000(1.12)^{10} = \$15{,}529.24$.
   (b) We need to find the value of t such that $5000(1.12)^t = 10{,}000$:

$$\begin{aligned} 5000(1.12)^t &= 10{,}000 \\ 1.12^t &= 2 && \text{dividing by 5000} \\ \log\left(1.12^t\right) &= \log 2 && \text{taking logs} \\ t \cdot \log 1.12 &= \log 2 && \text{using a log property} \\ t &= \frac{\log 2}{\log 1.12} && \text{dividing} \\ &= 6.12. \end{aligned}$$

This means it will take just over 6 years for the balance to reach \$10,000. We can use a similar approach to find out how long it takes the balance to reach \$20,000:

$$\begin{aligned} 5000(1.12)^t &= 20{,}000 \\ 1.12^t &= 4 && \text{dividing by 5000} \\ \log\left(1.12^t\right) &= \log 4 && \text{taking logs} \\ t \cdot \log 1.12 &= \log 4 && \text{using a log property} \\ t &= \frac{\log 4}{\log 1.12} && \text{dividing} \\ &= 12.2. \end{aligned}$$

This means it will take just over 12 years for the balance to reach \$20,000.

5. To match formula and graph, we keep in mind the effect on the graph of the parameters a and b in $y = ab^t$.
   If $a > 0$ and $b > 1$, then the function is positive and increasing.
   If $a > 0$ and $0 < b < 1$, then the function is positive and decreasing.
   If $a < 0$ and $b > 1$, then the function is negative and decreasing.
   If $a < 0$ and $0 < b < 1$, then the function is negative and increasing.
   (a) $y = 8.3e^{-t}$, so $a = 8.3$ and $b = e^{-1}$. Since $a > 0$ and $0 < b < 1$, we want a graph which is positive and decreasing. The graph in (ii) satisfies this condition.
   (b) $y = 2.5e^t$, so $a = 2.5$ and $b = e$. Since $a > 0$ and $b > 1$, we want a graph which is positive and increasing, such as (i).
   (c) $y = -4e^{-t}$, so $a = -4$ and $b = e^{-1}$. Since $a < 0$ and $0 < b < 1$, we want a graph which is negative and increasing, such as (iii).

9. Since this function is exponential, we know $y = ab^x$. We also know that $(-2, 8/9)$ and $(2, 9/2)$ are on the graph of this function, so

$$\frac{8}{9} = ab^{-2}$$

and

$$\frac{9}{2} = ab^2.$$

From these two equations, we can say that

$$\frac{\frac{9}{2}}{\frac{8}{9}} = \frac{ab^2}{ab^{-2}}.$$

Since $(9/2)/(8/9) = 9/2 \cdot 9/8 = 81/16$, we can re-write this equation to be

$$\frac{81}{16} = b^4.$$

Keeping in mind that $b > 0$, we get

$$b = \sqrt[4]{\frac{81}{16}} = \frac{\sqrt[4]{81}}{\sqrt[4]{16}} = \frac{3}{2}.$$

Substituting $b = 3/2$ in $ab^2 = 9/2$, we get

$$\frac{9}{2} = a(\frac{3}{2})^2 = \frac{9}{4}a$$

$$a = \frac{\frac{9}{2}}{\frac{9}{4}} = \frac{9}{2} \cdot \frac{4}{9} = \frac{4}{2} = 2.$$

Thus, $y = 2(3/2)^x$.

13. This equation cannot be solved analytically. Graphing $y = 87e^{0.066t}$ and $y = 3t + 7$ it is clear that these graphs will not intersect, which means $87e^{0.66t} = 3t + 7$ has no solution. The concavity of the graphs ensures that they will not intersect beyond the portions of the graphs shown in Figure 4.19.

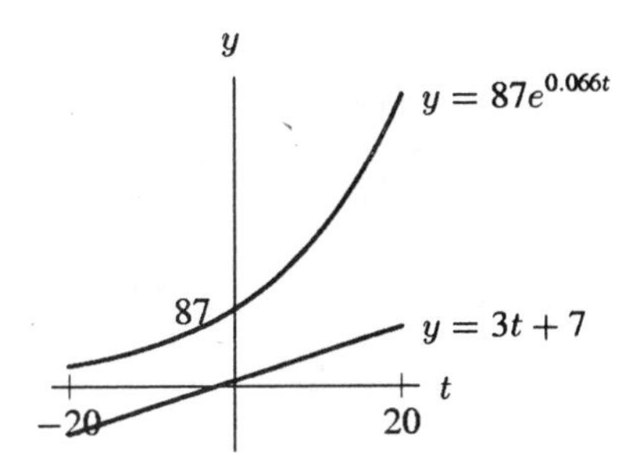

Figure 4.19

17. Notice that the x-values are not equally spaced. By finding $\Delta f/\Delta x$ in Table 4.11 we see $f(x)$ is linear, because the rates of change are constant.

TABLE 4.11

x	$f(x)$	$\frac{\Delta f(x)}{\Delta(x)}$
0.21	0.03193	
		0.093
0.37	0.04681	
		0.093
0.41	0.05053	
		0.093
0.62	0.07006	
		0.093
0.68	0.07564	

Since $f(x)$ is linear its formula will be $f(x) = b + mx$. From the table, we know $m = 0.093$. Choosing the point $(0.41, 0.05053)$, we have

$$\begin{aligned} 0.05053 &= b + 0.093(0.41) \\ b &= 0.0124. \end{aligned}$$

Thus, $f(x) = 0.0124 + 0.093x$.

Alternately, we could have tested $f(x)$ to see if it is exponential with a constant base b. Checking values will show that any proposed b would not remain constant when calculated with different points of function values for $f(x)$ from the table.

Since f is linear, we conclude that g is exponential. Thus, $g(x) = ab^x$. Using two points from above, we have

$$g(0.21) = ab^{0.21} = 3.324896 \qquad g(0.37) = ab^{0.37} = 3.423316.$$

So taking the ratios of $ab^{0.37}$ and ab^{021}, we have

$$\begin{aligned} \frac{ab^{0.37}}{ab^{0.21}} &= \frac{3.423316}{3.324896} \\ b^{0.16} &\approx 1.0296. \end{aligned}$$

So

$$b = (1.0296)^{\frac{1}{0.16}} \approx 1.20.$$

Substituting this value of B in the first equation gives

$$a(1.20)^{0.21} = 3.324896.$$

So

$$\begin{aligned} a &= \frac{3.324896}{1.20^{0.21}} \\ a &\approx 3.2. \end{aligned}$$

Thus, $g(x)$ is approximated by the formula

$$g(x) = 3.2(1.2)^x.$$

21. (a) For a linear model, we assume that the population increases by the same amount every year. Since it grew by 4.14% in the first year, the town had a population increase of $0.0414(20{,}000) = 828$ people in one year. According to a linear model, the population in 1990 would be $20{,}000 + 10 \cdot 828 = 28{,}280$. Using an exponential model, we assume that the population increases by the same percent every year, so the population in 1990 would be $20{,}000 \cdot (1.0414)^{10} \approx 30{,}006$. Clearly the exponential model is a better fit.

(b) Assuming exponential growth at 4.14% a year, the formula for the population is

$$P(t) = 20{,}000(1.0414)^t.$$

25. The annual growth factors for this investment are 1.27, 1.36, 1.19, 1.44, and 1.57. Thus, the investment increases by a total factor of $(1.27)(1.36)(1.19)(1.44)(1.57) \approx 4.6468$, indicating that the investment is

464.68% of what it had been. If x is the average annual growth factor for this five-year period, this means

$$x^5 = 4.6468$$

and so

$$\begin{aligned} x &= 4.6468^{1/5} \\ &\approx 1.3597. \end{aligned}$$

This tells us that, for each of the five years, the investment has, on average, 135.97% the value of the previous year. It is growing by 35.97% each year. Notice that if we had instead summed these percentages and divided the result by 5, we would have obtained a growth factor of 36.6%. This would be wrong because, as you can check for yourself, 5 years of 36.6% annual growth would result in a total increase of 475.6%, which is not correct.

29. (a) Because the time intervals are equally spaced at $t = 1$ units apart, we can estimate b by looking at the ratio of successive populations, namely,

$$\frac{0.901}{0.755} \approx 1.193, \quad \frac{1.078}{0.901} \approx 1.196, \quad \frac{1.285}{1.078} \approx 1.192.$$

These are nearly equal, the average being approximately 1.194, so $b \approx 1.194$. Using the data point $(0, 0.755)$ we estimate $a = 0.755$. Figure 4.20 shows the population data as well as the function $P = 0.755\,(1.194)^t$, where t is the number of 5-year intervals since 1975.

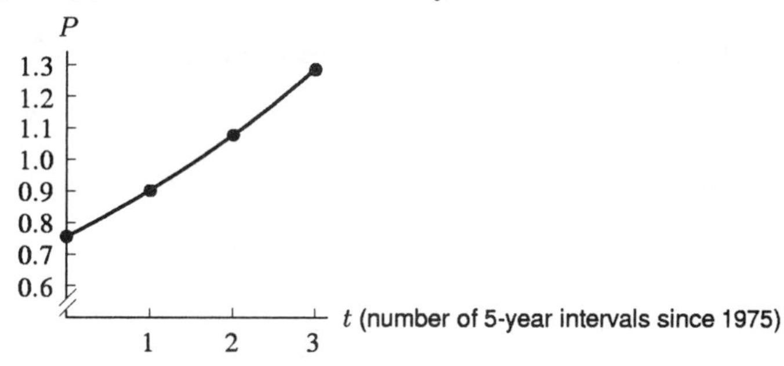

Figure 4.20: Actual and theoretical population of Botswana

(b) To find when the population doubles, we need to find the time t when $P = 2 \cdot 0.755$, that is, when $2 \cdot 0.755 = 0.755\,(1.194)^t$, or $2 = (1.194)^t$. By using a calculator to compute $(1.194)^t$ for $t = 1, 2, 3$, and so on, we find $(1.194)^4 \approx 2.03$. (Recall that t is measured in 5-year intervals.) This means the population doubles about every $5 \cdot 4 = 20$ years. Continuing in this way we see that $0.755\,(1.194)^{32} \approx 219.8$. Thus, according to this model, about $5 \cdot 32 = 160$ years from 1975, or 2135, the population of Botswana will exceed the 1975 population of the United States.

33. (a) We have $f(x) = 5(1.121)^x$ and we want it in the form $f(x) = ae^{kx}$. Evaluating $f(0)$ in both cases we have

$$\begin{aligned} f(0) &= 5(1.121)^0 = 5(1) = 5 \\ \text{and} \quad f(0) &= ae^{k \cdot 0} = ae^0 = a \end{aligned}$$

so we know that $a = 5$, and $f(x) = 5e^{kx}$. Equating the two expressions for $f(x)$, we have

$$\begin{aligned} 5e^{kx} &= 5(1.121)^x \\ e^{kx} &= (1.121)^x \\ e^k &= 1.121 \\ \ln e^k &= \ln 1.121 \\ k &= \ln 1.121 \approx 0.1142. \end{aligned}$$

So $f(x) = 5e^{0.1142x}$.

(b) We can rewrite

$$g(x) = 17e^{0.094x} = 17(e^{0.094})^x \approx 17(1.0986)^x,$$

with $a = 17$ and $b \approx 1.0986$.

(c) To get $h(x) = 22(2)^{\frac{x}{15}}$ in the form $h(x) = ab^x$, we just need to use the law of exponents:

$$h(x) = 22(2)^{\frac{x}{15}} = 22(2)^{\frac{1}{15}\cdot x} = 22(2^{1/15})^x \approx 22(1.0473)^x,$$

with $a = 22$ and $b \approx 1.0473$.

To rewrite $h(x) = 22(2)^{\frac{x}{15}}$ in the form $h(x) = ae^{kx}$, we evaluate $h(0)$ in both cases and find that $a = 22$. Equating $h(x) = 22(2)^{\frac{x}{15}}$ and $h(x) = 22e^{kx}$ we get:

$$\begin{aligned} 22(2)^{\frac{x}{15}} &= 22e^{kx} \\ 2^{\frac{x}{15}} &= e^{kx} \\ 2^{\frac{1}{15}} &= e^k \\ \ln e^k &= \ln 2^{\frac{1}{15}} = \frac{\ln 2}{15} \\ k &= \frac{\ln 2}{15} \approx 0.0462. \end{aligned}$$

Thus

$$h(x) = 22e^{0.0462x}.$$

37. (a) Using $B = 4.250$ and $T = 2.5$, $R = \log\left(\frac{a}{T}\right) + B$ becomes

$$R = \log\left(\frac{a}{2.5}\right) + 4.25.$$

If $R = 6.1$, then we want to solve

$$\begin{aligned} 6.1 &= \log\left(\frac{a}{2.5}\right) + 4.250 \\ 1.850 &= \log\left(\frac{a}{2.5}\right). \end{aligned}$$

If $y = \log x$, then $10^y = x$. So we can rewrite this equation to get

$$\begin{aligned} 10^{1.850} &= \frac{a}{2.5} \\ (2.5)(10^{1.850}) &= a \\ a &\approx (2.5)(70.8) \approx 177 \text{ microns.} \end{aligned}$$

(b) Another way to find the value of a is to first solve the equation for a

$$\begin{aligned} R &= \log\left(\frac{a}{T}\right) + B \\ R - B &= \log\left(\frac{a}{T}\right). \end{aligned}$$

Writing in exponential form:

$$\begin{aligned} 10^{(R-B)} &= \frac{a}{T} \\ a &= 10^{(R-B)}(T) \end{aligned}$$

In this case,

$$a = 10^{(7.1-4.250)}(2.5) \approx 1770 \text{ microns.}$$

(c) The values of R differ by 1 ($= 7.1 - 6.1$), but the values of a differ by a factor of 10 ($= 1770/177$).

41. (a) To find the average rate of change, we find the slope of the line connecting $(1950, 43)$ and $(1992, 15)$ on the graph of African-American infant mortality:

$$\text{Slope} = \frac{15 - 43}{1992 - 1950} = -0.67.$$

Similarly, for Caucasian infant mortality, we have

$$\text{Slope} = \frac{5 - 26}{1992 - 1950} = -0.5$$

It appears that African-American infant mortality declined faster on this interval.

(b) The ratio of African-American to Caucasian infant mortality in 1950 was $\frac{43}{26} \approx 1.65$. In 1992 it was $\frac{15}{5} = 3$. These ratios tell us that Caucasian infant deaths were a smaller fraction of African-American infant deaths in 1992 than in 1950. From this point of view, Caucasian infant mortality declined faster.

(c) Reading from the graph gives the infant mortality rates in Table 4.12.

TABLE 4.12

t (years since 1950)	African-American	Caucasian
0	43	26
5	42	23
10	43	22
15	40	21
20	30	17
25	26	13
30	20	10
35	17	8
40	16	6

(d) The ratios of successive terms are not constant, so these data do not exhibit exponential decline.

45. (a)

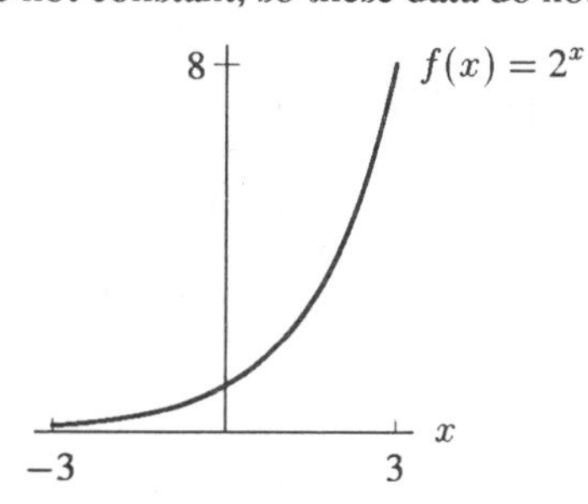

(b) The point $(0, 1)$ is on the graph. So is $(0.01, 1.00696)$. Taking $\frac{y_2 - y_1}{x_2 - x_1}$, we get an estimate for the slope of 0.696. We may zoom in still further to find that $(0.001, 1.000693)$ is on the graph. Using this and the point $(0, 1)$ we would get a slope of 0.693. Zooming in still further we find that the slope stabilizes at around 0.693; so, to two digits of accuracy, the slope is 0.69.

(c) Using the same method as in part (b), we find that the slope is ≈ 1.10.

(d) We might suppose that the slope of the tangent line at $x = 0$ increases as b increases. Trying a few values, we see that this is the case. Then we can find the correct b by trial and error: $b = 2.5$ has slope around 0.916, $b = 3$ has slope around 1.1, so $2.5 < b < 3$. Trying $b = 2.75$ we get a slope of 1.011, just a little too high. $b = 2.7$ gives a slope of 0.993, just a little too low. $b = 2.72$ gives a slope of 1.0006, which is as good as we can do by giving b to two decimal places. Thus $b \approx 2.72$.

In fact, the slope is exactly 1 when $b = e = 2.718\ldots$.

CHAPTER FIVE

Solutions for Section 5.1

1. (a)

x	-1	0	1	2	3
$g(x)$	-3	0	2	1	-1

The graph of $g(x)$ is shifted one unit to the right of $f(x)$.

(b)

x	-3	-2	-1	0	1
$h(x)$	-3	0	2	1	-1

The graph of $h(x)$ is shifted one unit to the left of $f(x)$.

(c)

x	-2	-1	0	1	2
$k(x)$	0	3	5	4	2

The graph $k(x)$ is shifted up three units from $f(x)$.

(d)

x	-1	0	1	2	3
$m(x)$	0	3	5	4	2

The graph $m(x)$ is shifted one unit to the right and three units up from $f(x)$.

5. (a) $k(w) - 3 = 3^w - 3$
To sketch, shift the graph of $k(w) = 3^w$ down 3 units, as in Figure 5.1.

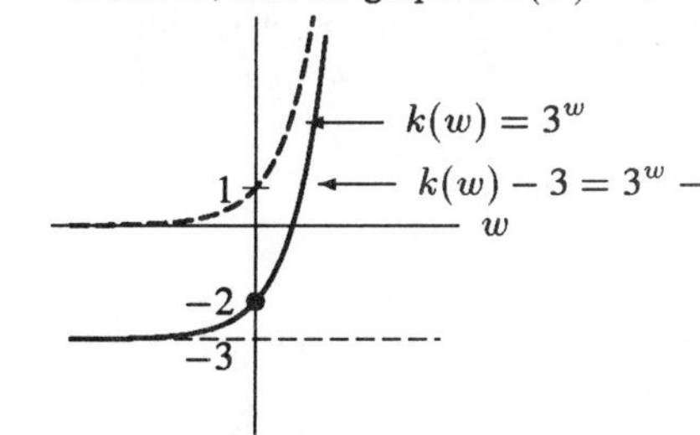

Figure 5.1

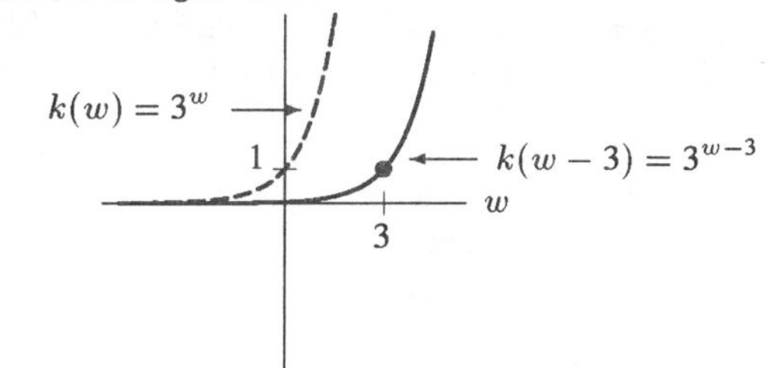

Figure 5.2

(b) $k(w - 3) = 3^{w-3}$
To sketch, shift the graph of $k(w) = 3^w$ to the right by 3 units, as in Figure 5.2.

(c) $k(w) + 1.8 = 3^w + 1.8$
To sketch, shift the graph of $k(w) = 3^w$ up by 1.8 units , as in Figure 5.3.

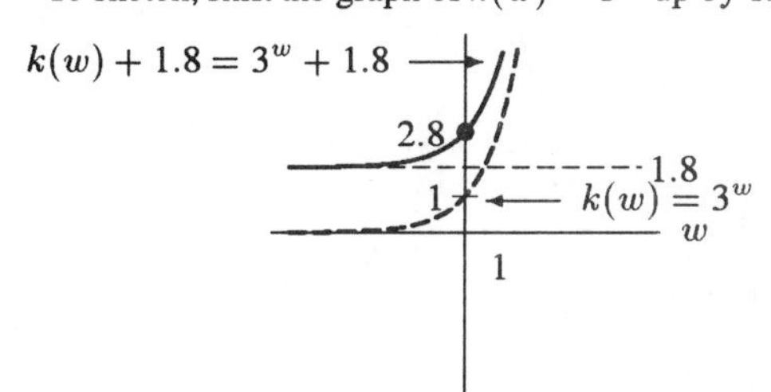

Figure 5.3

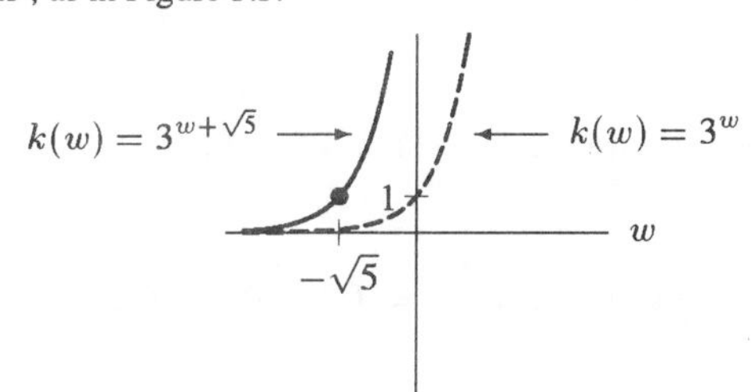

Figure 5.4

(d) $k(w + \sqrt{5}) = 3^{w+\sqrt{5}}$
To sketch, shift the graph of $k(w) = 3^w$ to the left by $\sqrt{5}$ units, as in Figure 5.4.

(e) $k(w+2.1)-1.3=3^{w+2.1}-1.3$
To sketch, shift the graph of $k(w)=3^w$ to the left by 2.1 units and down 1.3 units, as in Figure 5.5.

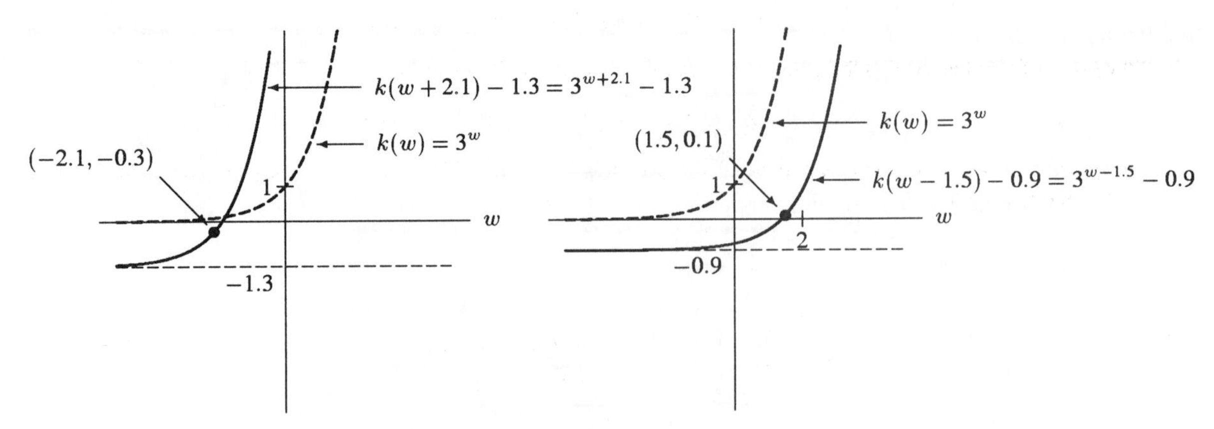

Figure 5.5 ***Figure 5.6***

(f) $k(w-1.5)-0.9=3^{w-1.5}-0.9$
To sketch, shift the graph of $k(w)=3^w$ to the right by 1.5 units and down by 0.9 units, as in Figure 5.6.

9. (a) This is an outside change, and thus a vertical change, to $y=|x|$. The graph of $g(x)$ is the graph of $|x|$ shifted upward by 1 unit. See Figure 5.7.
   (b) This is an inside change, and thus a horizontal change, to $y=|x|$. The graph of $h(x)$ is the graph of $|x|$ shifted to the left by 1 unit. See Figure 5.8.
   (c) The graph of $j(x)$ involves two transformations of the graph of $y=|x|$. First, the graph is shifted to the right by 2 units. Next, the graph is shifted up by 3 units. Figure 5.9 shows the result of these two consecutive transformations.

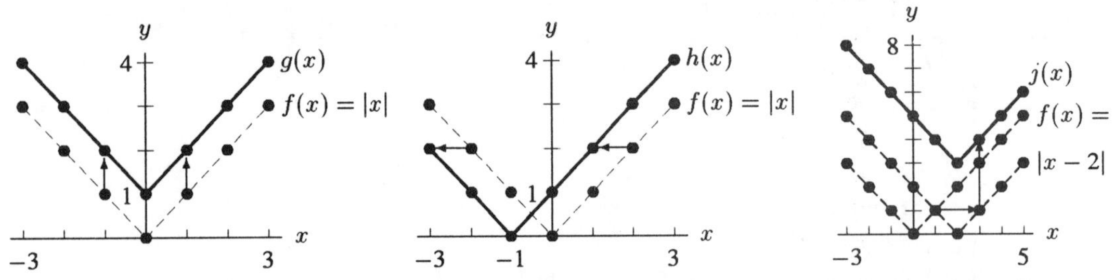

Figure 5.7: Graph of $g(x)=|x|+1$, the graph of $|x|$ shifted up 1 unit

Figure 5.8: Graph of $h(x)=|x+1|$, the graph of $|x|$ shifted left 1 unit

Figure 5.9: Graph of $j(x)=|x-2|+3$, the graph of $|x|$ shifted right 2 units and up 3 units

13. (a) Notice that the value of $h(x)$ at every value of x is 2 less than the value of $f(x)$ at the same x value. Thus

$$h(x)=f(x)-2.$$

   (b) Observe that $g(0)=f(1), g(1)=f(2)$, and so on.In general,

$$g(x)=f(x+1).$$

   (c) The values of $i(x)$ are two less than the values of $g(x)$at the same x value. Thus

$$i(x)=f(x+1)-2.$$

17. The graphs in Figure 5.10 appear to be vertical shifts of each other. The explanation for this relies on the property of logs which says that $\log(ab) = \log a + \log b$. Since

$$y = \log(10x) = \log 10 + \log x = 1 + \log x,$$

the graph of $y = \log(10x)$ is the graph of $y = \log x$ shifted up 1 unit. Similarly,

$$y = log(100x) = \log 100 + \log x = 2 + \log x,$$

the graph of $y = \log(100x)$ is the graph of $y = \log x$ shifted up 2 units. Thus, we see that the graphs are indeed vertical shifts of one another.

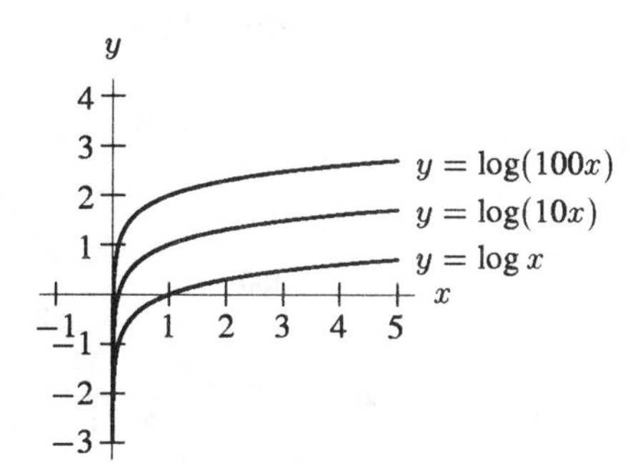

Figure 5.10

21. (a) There are many possible graphs, but all should show seasonally-related cycles of temperature increases and decreases.

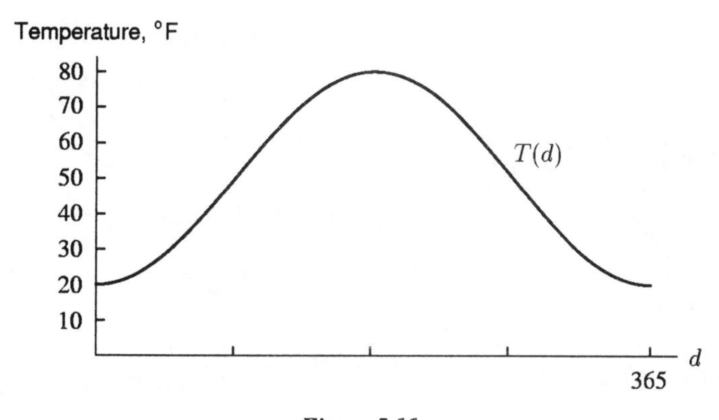

Figure 5.11

(b) While there are a wide variety of correct answers, the value of $T(6)$ is a temperature for a day in early January, $T(100)$ for a day in mid-April, and $T(215)$ for a day in early August. The value for $T(371) = T(365 + 6)$ should be close to that of $T(6)$.

(c) Since there are usually 365 days in a year, $T(d)$ and $T(d + 365)$ represent average temperatures on days which are a year apart.

(d) $T(d + 365)$ is the average temperature on the same day of the year a year earlier. They should be about the same value. Therefore, the graph of $T(d + 365)$ should be about the same as that of $T(d)$.

(e) The graph of $T(d) + 365$ is a shift upward of $T(d)$, by 365 units. It has no significance in practical terms, other than to represent a temperature that is 365° hotter than the average temperature on day d.

25. (a) If your taxable income is between \$0 and \$20,000, your tax is 15% of that income. The graph of this relationship is a line segment lying on the line which contains the origin and the point $(20000, 3000)$ and which has a slope of 0.15. If your taxable income is between \$20,000 and \$49,000, your tax is \$3000 plus 28% of the income which is above \$20,000. The graph of this relationship is also a line segment. It lies on the line which contains the points $(20000, 3000)$ and $(49000, 11120)$ and which has slope of 0.28.

If your taxable income is above \$49,000, your tax is \$11,120 plus 31% of the income which is above \$49,000. The graph of this relationship is a ray. It lies on the line which contains the point $(49000, 11120)$ and has slope of 0.31. Figure 5.12 shows the graph of $I(d)$. On the graph note that the slopes of the last two pieces are so close (.28 versus 0.31) that is almost impossible to see the change in steepness that occurs at $(49000, 11120)$. See Figure 5.12.

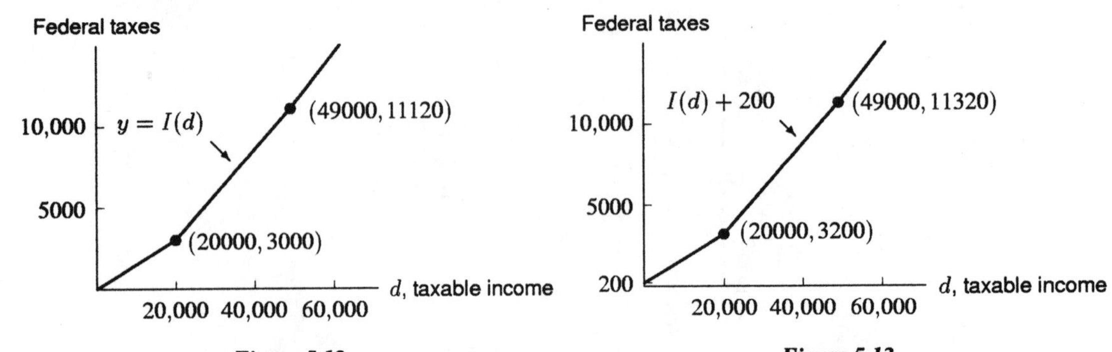

Figure 5.12 *Figure 5.13*

(b) Shift the graph of $I(d)$ upward by 200 units to get the graph of $I(d) + 200$ as shown in Figure 5.13. You will now owe \$200 more than under the old system defined by $I(d)$. See Figure 5.13.

(c) Slide the graph of $I(d)$ to the left by 1000 units to get the graph of $I(d + 1000)$, as shown in Figure 5.14. To get your new tax, add \$1000 to your taxable income and compute your tax under the old formulas. $I(d + 1000)$ might correspond to eliminating \$1000 of deductions.

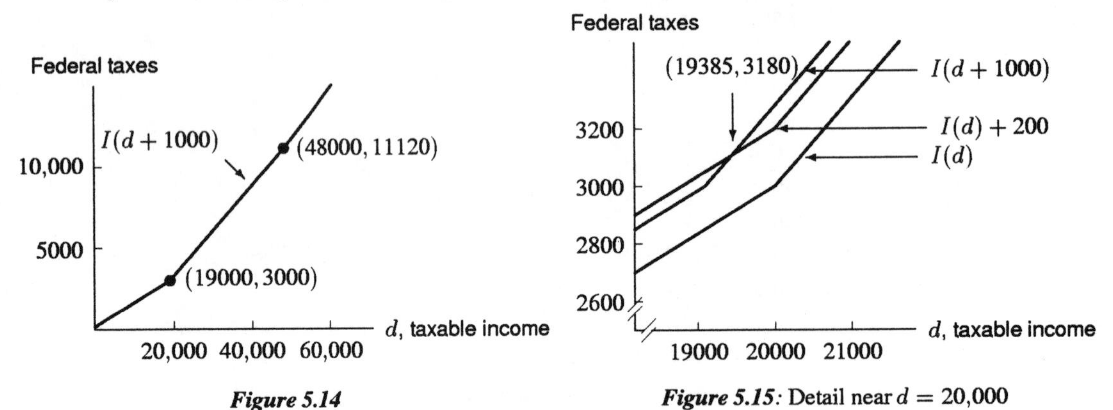

Figure 5.14 *Figure 5.15:* Detail near $d = 20{,}000$

(d) Under $I(d) + 200$, calculate your taxes using the old system, then add \$200. So, if your income is \$15,000,

$$I(15000) = 0.15(15000) = \$2250$$

and

$$I(15000) + 200 = 22500 + 200 = \$2450.$$

Under $I(d + 1000)$, you must add \$1000 to your taxable income and then calculate your taxes. So, if your income is \$15,000, then

$$I(d + 1000) = I(15000 + 1000) = I(16000) = 0.15(16000) = \$2400.$$

You would probably prefer $I(d + 1000)$ because you pay \$50.00 less.

(e) If income $d = 30{,}000$, then

$$I(30{,}000) + 200 = 3000 + 0.28(10{,}000) + 200 = \$6000$$
$$I(30{,}000 + 1000) = 3000 + 0.28(11{,}000) = \$6080.$$

So the $I(x) + 200$ is a better deal.

(f) It is probably easiest to determine when the tax obligations will be the same by comparing the graphs of $I(d) + 200$ and $I(d + 1000)$. In Figure 5.15, you can see a section of these graphs near $d = 20{,}000$. It is clear that the "break even point" is between 19,000 and 20,000. Now that you know this, you can use: a) the trace option on the graphs, b) a spreadsheet, or c) a system of equations to find the value of d for which the values of $I(d)$ and $I(d + 1000)$ are the same. If you want to solve this problem symbolically, use the fact that d is between 19,000 and 20,000 to choose the correct expression for calculating your taxes. Since

$$d \leq 20,000, \quad I(d) = 0.15d$$

so

$$I(d) + 200 = 0.15d + 200.$$

Also $d > 19{,}000$, so $d + 1000 > 20{,}000$. Therefore,

$$\begin{aligned} I(d + 1000) &= 3000 + 0.28((d + 1000) - 20{,}000) \\ &= 3000 + 0.28(d - 19{,}000) \\ &= 3000 + 0.28d - 5320 \\ &= 0.28d - 2320. \end{aligned}$$

Since we are looking for the value of d where

$$I(d) + 200 = I(d + 1000),$$

we can say that

$$\begin{aligned} 0.15d + 200 &= 0.28d - 2320 \\ 2520 &= 0.13d \\ d &\approx 19385. \end{aligned}$$

So, if you earn \$19,385, you will pay the same amount of tax, \$3108, whether you use $I(d) + 200$ or $I(d + 1000)$.

Solutions for Section 5.2

1.

TABLE 5.1

p	−3	−2	−1	0	1	2	3
$f(p)$	0	−3	−4	−3	0	5	12

TABLE 5.2

p	−3	−2	−1	0	1	2	3
$g(p)$	12	5	0	−3	−4	−3	0

TABLE 5.3

p	−3	−2	−1	0	1	2	3
$h(p)$	0	3	4	3	0	−5	−12

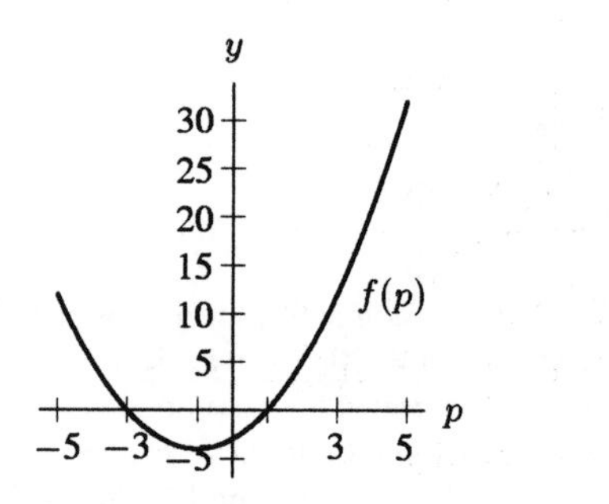

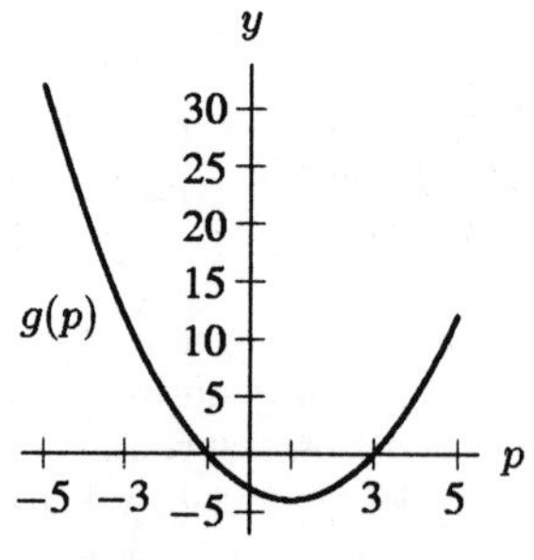

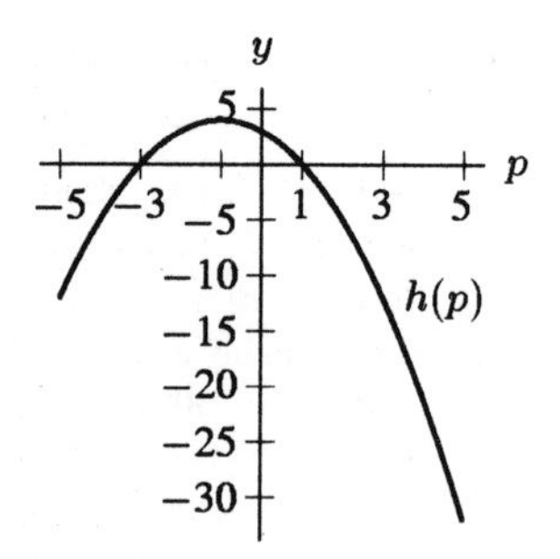

Figure 5.16: Graphs of $f(p)$, $g(p)$, and $h(p)$

Since $g(p) = f(-p)$, the graph of g is a horizontal reflection of the graph of f across the y-axis. Since $h(p) = -f(p)$, the graph of h is a reflection of the graph of f across the p-axis.

5.

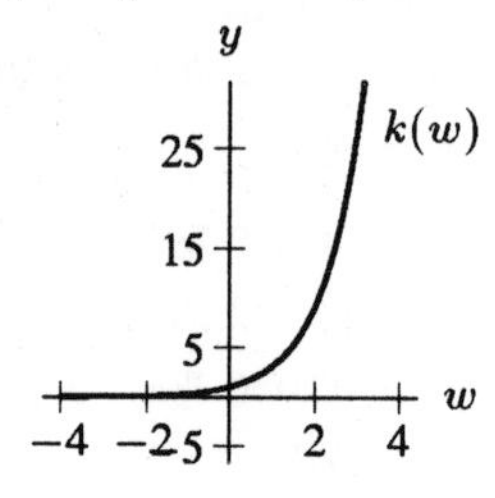

Figure 5.17: $k(w) = 3^w$

(a)

$$y = k(-w) = 3^{-w}$$

To graph this function, reflect the graph of k across the y-axis.

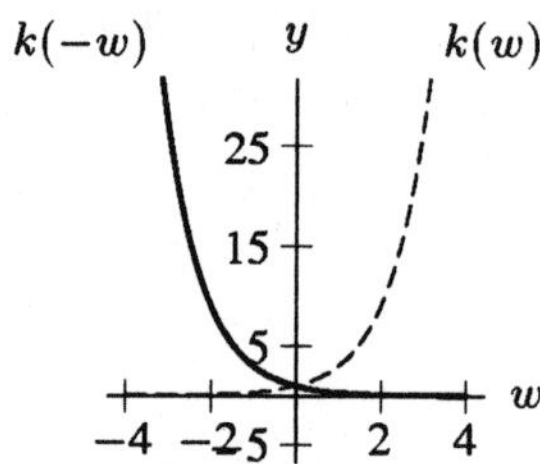

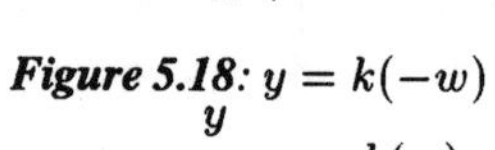

Figure 5.18: $y = k(-w)$

(b)

$$y = -k(w) = -3^w$$

To graph this function, reflect the graph of k across the w-axis.

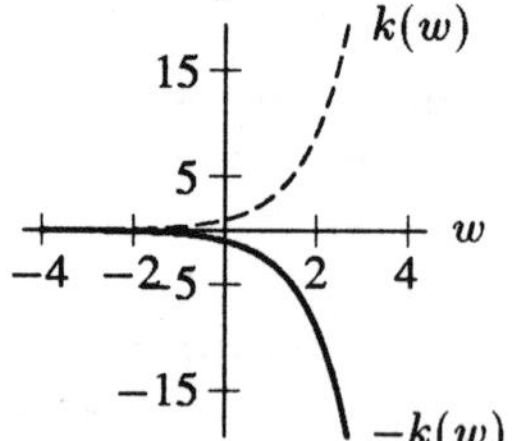

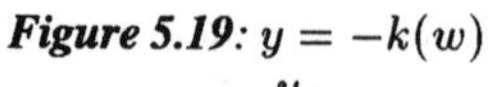

Figure 5.19: $y = -k(w)$

(c)

$$y = -k(-w) = -3^{-w}$$

To graph this function, first reflect the graph of k across the y-axis, then reflect it again across the w-axis.

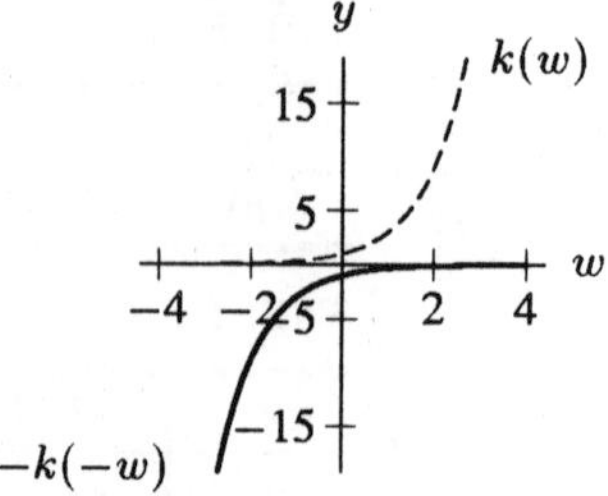

Figure 5.20: $y = -k(-w)$

(d)

$$y = -k(w-2) = -3^{w-2}$$

To graph this function, first reflect the graph of k across the w-axis, then shift it to the right by 2 units.

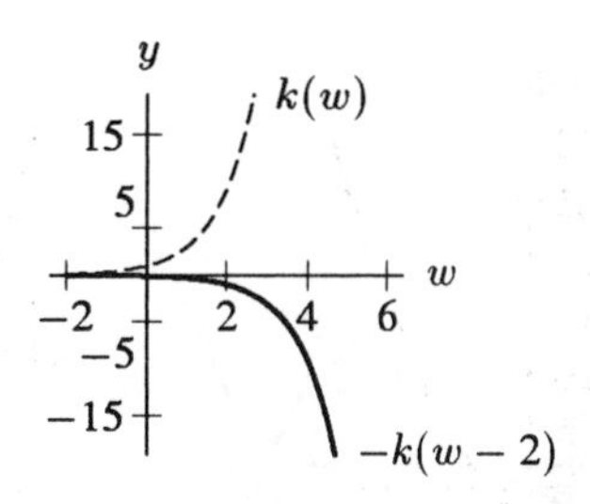

Figure 5.21: $y = -k(w-2)$

(e)

$$y = k(-w) + 4 = 3^{-w} + 4$$

To graph this function, first reflect the graph of k across the y-axis, then shift it up by 4 units.

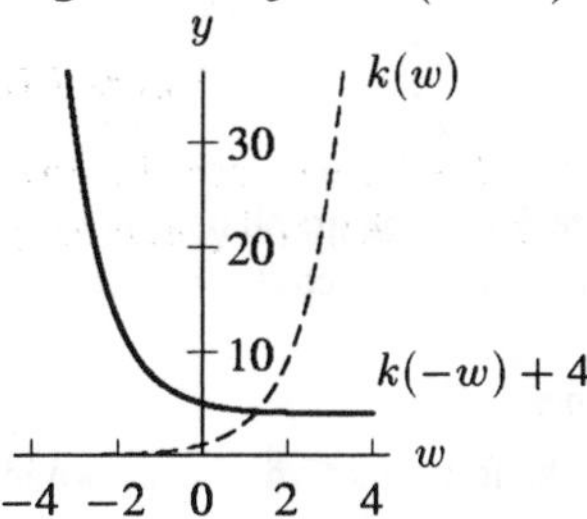

Figure 5.22: $y = k(-w) + 4$

(f)

$$y = -k(-w) - 1 = -3^{-w} - 1$$

To graph this function, first reflect the graph of k across the y-axis, then reflect it across the w-axis, finally shift it down by 1 unit.

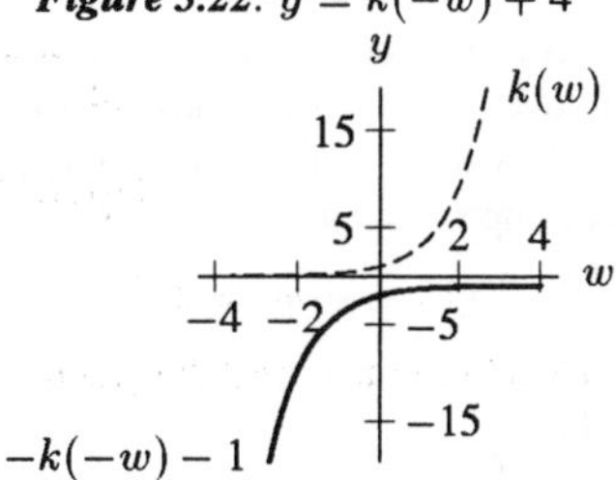

Figure 5.23: $y = -k(-w) - 1$

(g)

$$y = -3 - k(w) = -3 - 3^{w}$$

To graph this function, reflect the graph of k across the the w-axis and then shift it down 3 units.

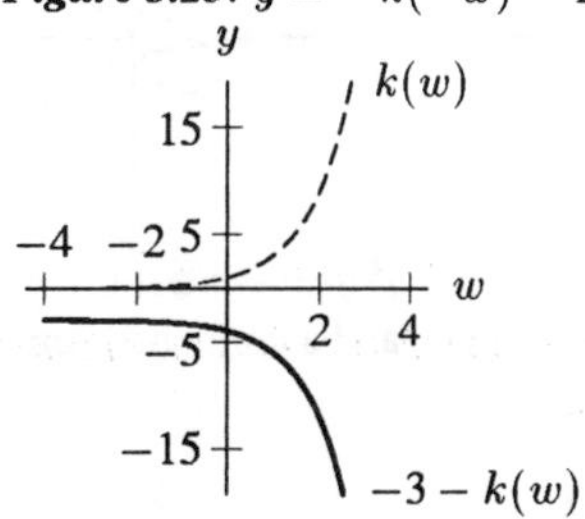

Figure 5.24: $y = -3 - k(w)$

9. (a) See Figure 5.25.

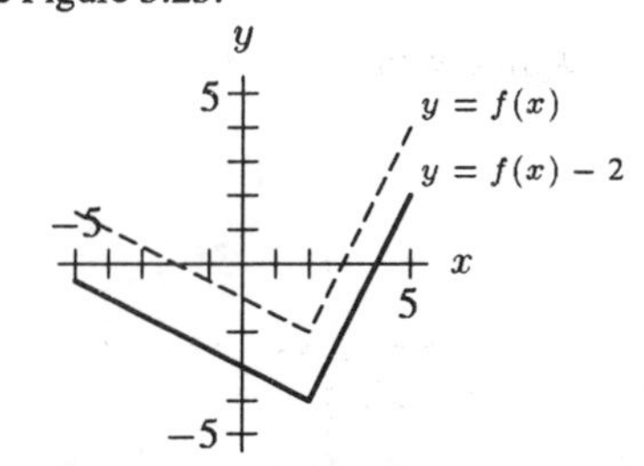

Figure 5.25: The graph of $f(x) - 2$ is the graph of $f(x)$ shifted down 2 units

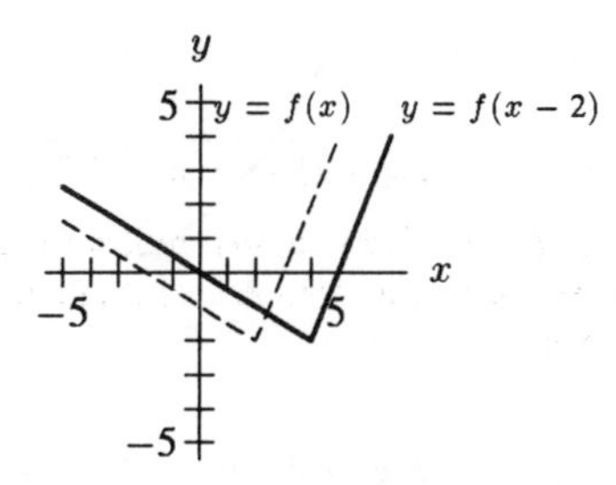

Figure 5.26: The graph of $f(x-2)$ is the graph of $f(x)$ shifted right 2 units

(b) See Figure 5.26.

(c) See Figure 5.27.

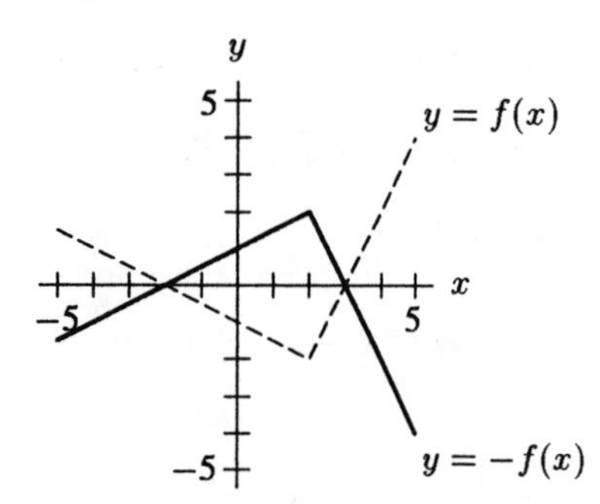

Figure 5.27: The graph of $-f(x)$ is the graph of $f(x)$ reflected across the x-axis

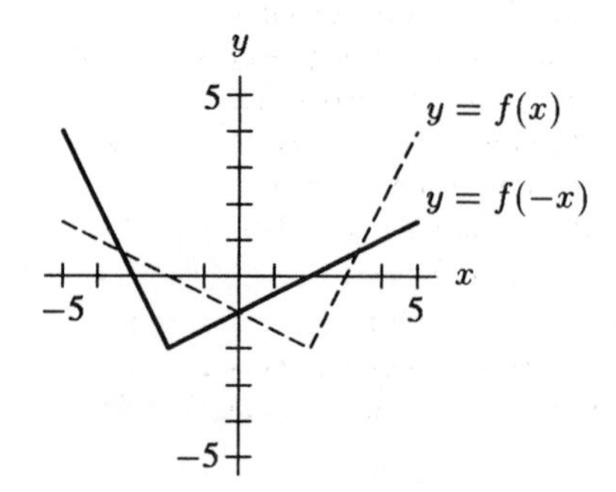

Figure 5.28: The graph of $f(-x)$ is the graph of $f(x)$ reflected across the y-axis

(d) See Figure 5.28.

13. The graphs of the four functions are shown in Figure 5.29. The graphs of $y = -g(x)$, $y = g(-x)$ and $y = -g(-x)$ are reflections of the graph of $y = g(x)$: $y = -g(x)$ is a reflection across the x-axis, $y = g(-x)$ is a reflection across the y-axis and $y = -g(-x)$ is a reflection across both axes. The graphs of $y = g(x)$ and $y = -g(-x)$ are both increasing, while the graphs of $y = g(-x)$ and $y = -g(x)$ are both decreasing. The graphs of $y = g(x)$ and $y = g(-x)$ are both concave up, above the x-axis and have a y-intercept at $(0, 3)$, while those of $y = -g(x)$ and $y = -g(-x)$ are both concave down, below the x-axis and have a y-intercept of $(0, -3)$. All four graphs approach the x-axis, but $y = g(x)$ and $y = -g(x)$ approach it as x becomes more and more negative, while $y = g(-x)$ and $y = -g(-x)$ approach it as x becomes larger.

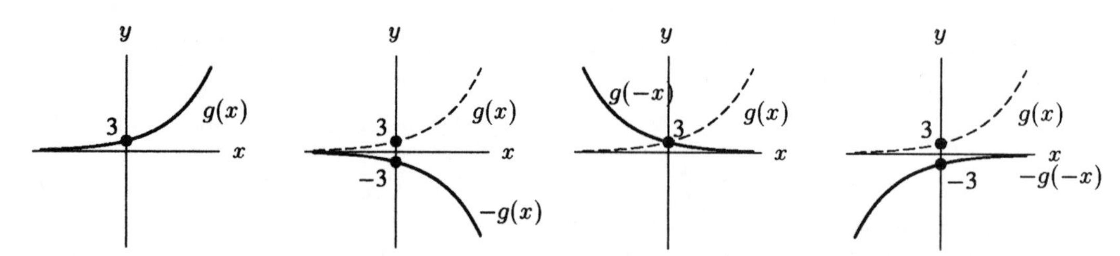

Figure 5.29: The graphs of $y = -g(x)$, $y = g(-x)$, and $y = -g(-x)$ are, respectively, a vertical flip, a horizontal flip, and a combined vertical and horizontal flip of the graph of $y = g(x)$

17.

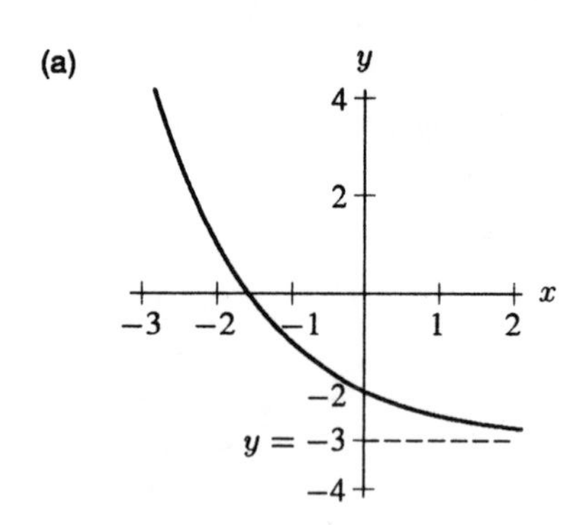

Figure 5.30: $y = 2^{-x} - 3$

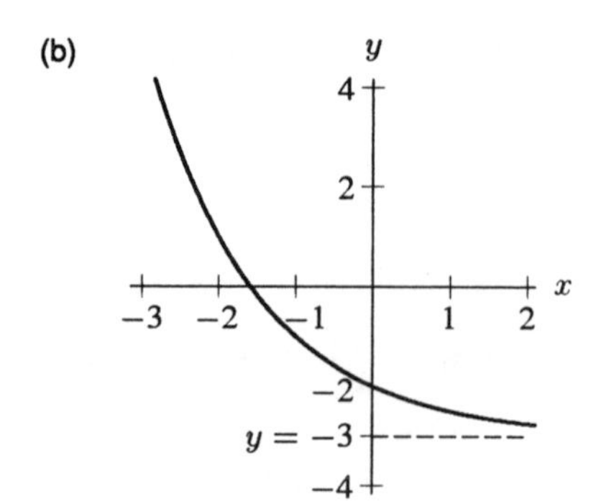

Figure 5.31: $y = 2^{-x} - 3$

(c) The two functions are the same in this case. Note that you will not always obtain the same result if you change the order of the transformations.

21. (a) Since the values of $f(x)$ and $f(-x)$ are the same, $f(x)$ appears to be symmetric across the y-axis. Thus, $f(x)$ could be an even function.
    (b) Since the value of $g(-x)$ is the opposite of $g(x)$, we know that $g(x)$ could be symmetric about the origin. Thus, $g(x)$ could be an odd function.
    (c) The value of $h(-x) = f(-x)+g(-x)$ is not the same as either $h(x) = f(x)+g(x)$ or $-h(x) = -(f(x)+g(x))$, so $h(x)$ is not symmetric.
    (d) Note that $j(3) = f(3+1) = f(4) = 13$ and $j(-3) = f(-3+1) = f(-2) = 1$. Thus, $j(-x)$ does not equal either $j(x)$ or $-j(x)$, so $j(x)$ is not symmetric.

25. One way to do this is to sketch a graph of $y = h(x)$ to see that it appears to be symmetric across the origin. In other words, we can visually check to see that flipping the graph of $y = h(x)$ about the y-axis and then the x-axis (or vice-versa) does not change the appearance of the function's graph. See Figure 5.32.

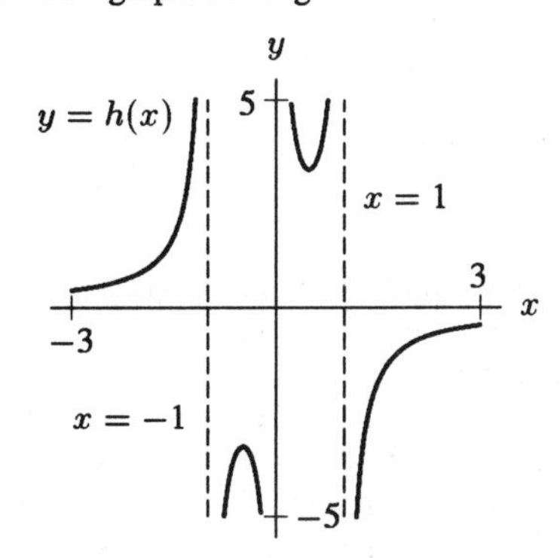

Figure 5.32: The graph of $y = h(x) = \frac{1+x^2}{x-x^3}$ is symmetric across the origin

To confirm that $h(x)$ is symmetric across the origin, we use algebra. We need to show that $h(-x) = -h(x)$ for any x. Finding the formula for $h(-x)$, we have

$$\begin{aligned} h(-x) &= \frac{1+(-x)^2}{(-x)-(-x)^3} = \frac{1+x^2}{-x+x^3} = \frac{1+x^2}{-(x-x^3)} \\ &= \frac{1+x^2}{x-x^3} = -h(x). \end{aligned}$$

Thus, the formula for $h(-x)$ is the same as the formula for $-h(x)$, and so the graph of $y = h(x)$ is symmetric across the origin.

29. Because $f(x)$ is an odd function, $f(x) = -f(-x)$. Setting $x = 0$ gives $f(0) = -f(0)$, so $f(0) = 0$. Since $c(0) = 1$, $c(x)$ is not odd. Since $d(0) = 1$, $d(x)$ is not odd.

33. (a) In order for $f(x)$ to be even,

$$\begin{aligned} f(-x) &= f(x) \\ m(-x)+b &= mx+b \\ -mx+b &= mx+b \text{ for all } x. \end{aligned}$$

This is true if and only if $-m = m$, which is true if and only if $m = 0$, so $f(x) = b$. Thus, a linear function is even only when it is a constant; its graph is a horizontal line.

    (b) In order for $f(x)$ to be odd,

$$\begin{aligned} f(-x) &= -f(x) \\ m(-x)+b &= -(mx+b) \\ -mx+b &= -mx-b \text{ for all } x. \end{aligned}$$

This is true if and only if $-b = b$, which is true if and only if $b = 0$, so $f(x) = mx$.

    (c) If $f(x)$ is both even and odd, then both (a) and (b) are true, which means $m = 0$ and $b = 0$, so $f(x) = 0$. The function $f(x) = 0$ is both even and odd, and its graph is the line $y = 0$, or the x-axis.

Solutions for Section 5.3

1. (a) To get the table for $f(x)/2$, you need to divide each entry for $f(x)$ by 2. See Table 5.4.

TABLE 5.4

x	-3	-2	-1	0	1	2	3
$f(x)/2$	1	1.5	3.5	-0.5	-1.5	2	4

(b) In order to get the table for $-2f(x+1)$, first get the table for $f(x+1)$. To do this, note that, if $x = 0$, then $f(x+1) = f(0+1) = f(1) = -3$ and if $x = -4$, then $f(x+1) = f(-4+1) = f(-3) = 2$. Since $f(x)$ is defined for $-3 \leq x \leq 3$, where x is an integer, then $f(x+1)$ is defined for $-4 \leq x \leq 2$.

TABLE 5.5

x	-4	-3	-2	-1	0	1	2
$f(x+1)$	2	3	7	-1	-3	4	8

Next, multiply each value of $f(x+1)$ entry by -2.

TABLE 5.6

x	-4	-3	-2	-1	0	1	2
$-2f(x+1)$	-4	-6	-14	2	6	-8	-16

(c) To get the table for $f(x)+5$, you need to add 5 to each entry for $f(x)$ in in the table given in the problem.

TABLE 5.7

x	-3	-2	-1	0	1	2	3
$f(x)+5$	7	8	12	4	2	9	13

(d) If $x = 3$, then $f(x-2) = f(3-2) = f(1) = -3$. Similarly if $x = 2$ then $f(x-2) = f(0) = -1$, since $f(x)$ is defined for integral values of x from -3 to 3, $f(x-2)$ is defined for integral values of x, which are two units higher, that is from -1 to 5.

TABLE 5.8

x	-1	0	1	2	3	4	5
$f(x-2)$	2	3	7	-1	-3	4	8

(e) If $x = 3$, then $f(-x) = f(-3) = 2$, whereas if $x = -3$, then $f(-x) = f(3) = 8$. So, to complete the table for $f(-x)$, flip the values of $f(x)$ given in the problem about the origin.

TABLE 5.9

x	-3	-2	-1	0	1	2	3
$f(-x)$	8	4	-3	-1	7	3	2

(f) To get the table for $-f(x)$, take the negative of each value of $f(x)$ from the table given in the problem.

TABLE 5.10

x	-3	-2	-1	0	1	2	3
$-f(x)$	-2	-3	-7	1	3	-4	-8

5. (a) Since $y = -f(x) + 2$, we first need to reflect the graph of $y = f(x)$ over the x-axis and then shift it upward two units. See Figure 5.33.

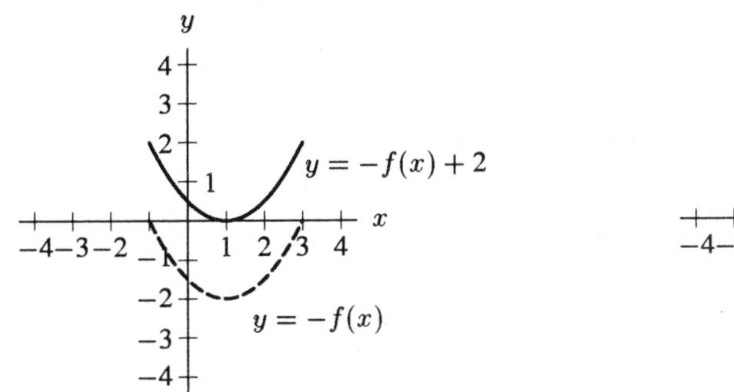

Figure 5.33

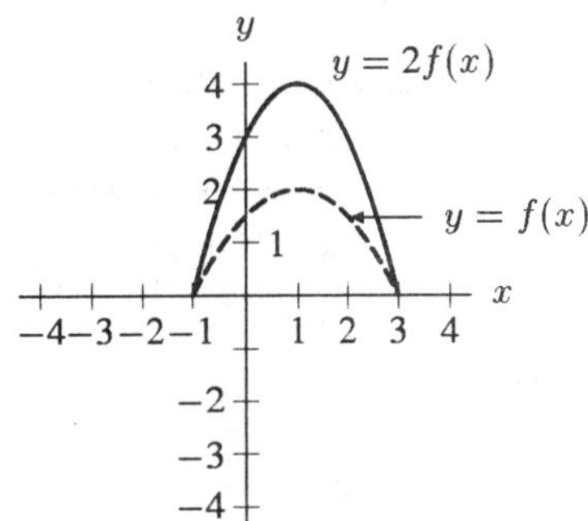

Figure 5.34

(b) We need to stretch the graph of $y = f(x)$ vertically by a factor of 2 in order to get the graph of $y = 2f(x)$. See Figure 5.34.

(c) In order to get the graph of $y = f(x - 3)$, we will move the graph of $y = f(x)$ to the right by 3 units. See Figure 5.35.

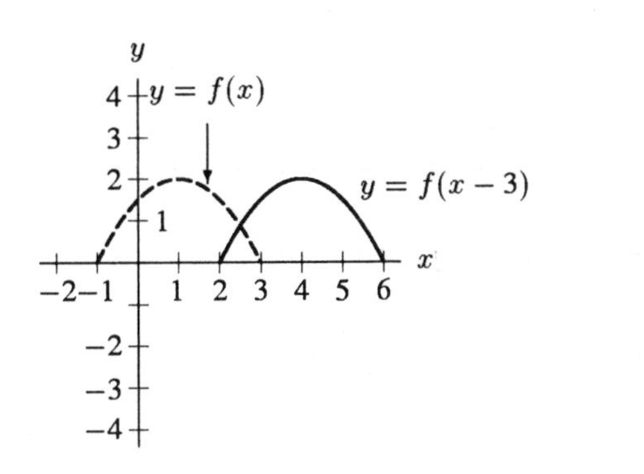

Figure 5.35

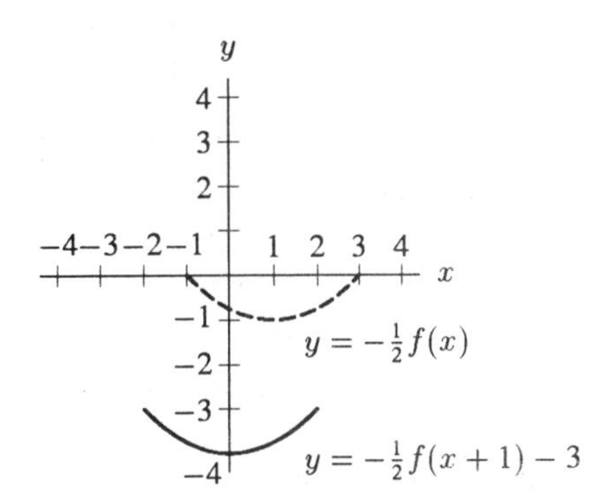

Figure 5.36

(d) To get the graph of $y = -\frac{1}{2}f(x + 1) - 3$, first move the graph of $y = f(x)$ to the left 1 unit, then compress it vertically by a factor of 2. Reflect this new graph over the x-axis and then move the graph down 3 units. See Figure 5.36.

9. (i) i: The graph of $y = f(x)$ has been stretched vertically by a factor of 2.
(ii) c: The graph of $y = f(x)$ has been stretched vertically by $1/3$, or compressed.
(iii) b: The graph of $y = f(x)$ has been reflected over the x-axis and raised by 1.
(iv) g: The graph of $y = f(x)$ has been shifted left by 2, and raised by 1.
(v) d: The graph of $y = f(x)$ has been reflected over the y-axis.

13. (a) This figure is the graph of $f(t)$ shifted upwards by two units. Thus its formula is $y = f(t) + 2$. Since on the graph of $f(t)$ the asymptote occurs at $y = 5$ on this graph the asymptote must occur at $y = 7$.
(b) This figure is the graph of $f(t)$ shifted to the left by one unit. Thus its formula is $y = f(t + 1)$. Since on the graph of $f(t)$ the asymptote occurs at $y = 5$, on this graph the asymptote also occurs at $y = 5$. Note that a horizontal shift does not affect the horizontal asymptotes.
(c) This figure is the graph of $f(t)$ shifted downwards by three units and to the right by two units. Thus its formula is $y = f(t - 2) - 3$. Since on the graph of $f(t)$ the asymptote occurs at $y = 5$, on this graph the asymptote must occur at $y = 2$. Again, the horizontal shift does not affect the horizontal asymptote. However, outside changes (vertical shifts) do change the horizontal asymptote.

17. Figure 5.37 gives a graph of a function $y = f(x)$ together with graphs of $y = \frac{1}{2}f(x)$ and $y = 2f(x)$. All three graphs cross the x-axis at $x = -2, x = -1$, and $x = 1$. Likewise, all three functions are increasing and decreasing on the same intervals. Specifically, all three functions are increasing for $x < -1.55$ and for $x > 0.21$ and decreasing for $-1.55 < x < 0.21$.

Even though the stretched and compressed versions of f shown by Figure 5.37 are increasing and decreasing on the same intervals, they are doing so at different rates. You can see this by noticing that, on every interval of x, the graph of $y = \frac{1}{2}f(x)$ is less steep than the graph of $y = f(x)$. Similarly, the graph of $y = 2f(x)$ is steeper than the graph of $y = f(x)$. This indicates that the magnitude of the average rate of change of $y = \frac{1}{2}f(x)$ is less than that of $y = f(x)$, and that the magnitude of the average rate of change of $y = 2f(x)$ is greater than that of $y = f(x)$.

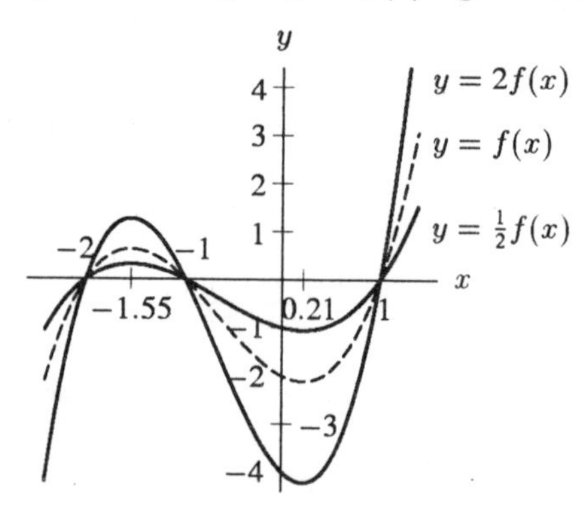

Figure 5.37: The graph of $y = 2f(x)$ and $y = \frac{1}{2}f(x)$ compared to the graph of $f(x)$

Solutions for Section 5.4

1. If $x = -2$, then $f(\frac{1}{2}x) = f(\frac{1}{2}(-2)) = f(-1) = 7$, and if $x = 6$, then $f(\frac{1}{2}x) = f(\frac{1}{2} \cdot 6) = f(3) = 8$. In general, $f(\frac{1}{2}x)$ is defined for values of x which are twice the values for which $f(x)$ is defined.

TABLE 5.11

x	-6	-4	-2	0	2	4	6
$f(\frac{1}{2}x)$	2	3	7	-1	-3	4	8

5.

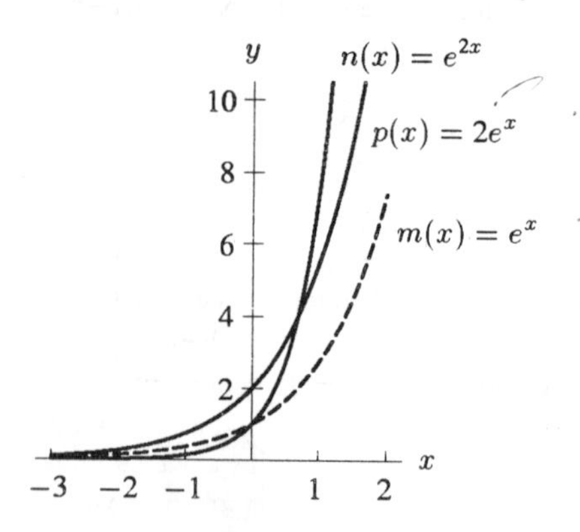

Figure 5.38

The graph of $n(x) = e^{2x}$ is a horizontal compression of the graph of $m(x) = e^x$. The graph of $p(x) = 2e^x$ is a vertical stretch of the graph of $m(x) = e^x$. All three graphs have a horizontal asymptote at $y = 0$. The y-intercept of $n(x) = e^{2x}$ is the same as for $m(x)$, but the graph of $p(x) = 2e^x$ has a y-intercept of $(0, 2)$.

9. The graph of $y = f(2x)$ is a horizontal compression of the graph of $y = f(x)$ by a factor of 2. The graph of $y = f(-\frac{x}{3}) = f(-\frac{1}{3}x)$ is both a horizontal stretch by a factor of 3 and a flip across the y-axis.

(a)

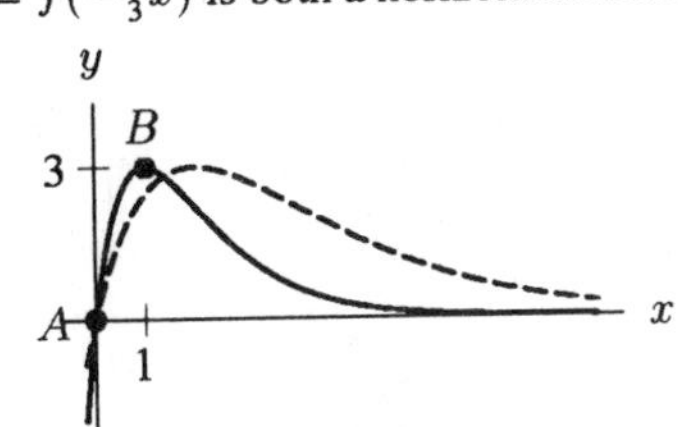

(b)

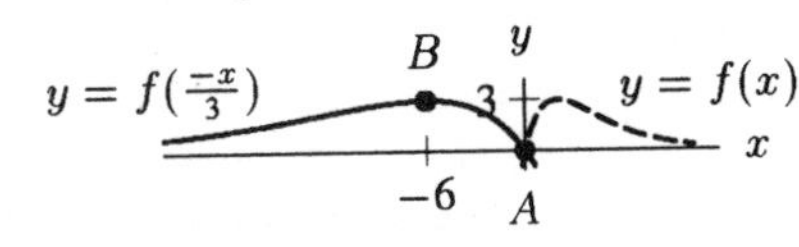

13. (a) Graphing f and g shows that there is a vertical shift of $+1$. See Figure 5.39.

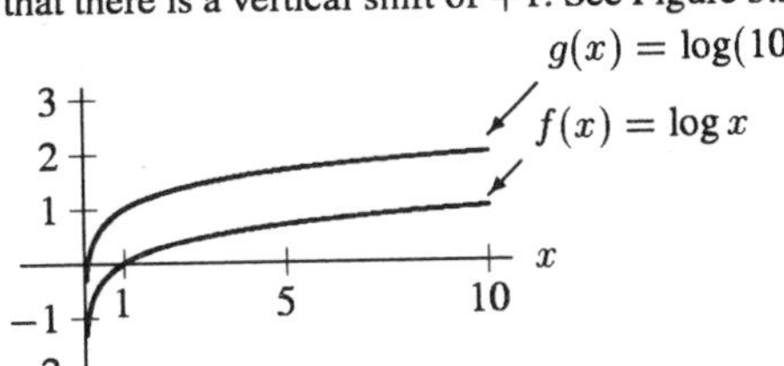

Figure 5.39: A vertical shift of $+1$

(b) Using the property that $\log(ab) = \log a + \log b$, we have

$$g(x) = \log(10x) = \log 10 + \log x = 1 + f(x).$$

Thus, $g(x)$ is $f(x)$ shifted vertically upward by 1.

(c) Using the same property of logarithms

$$\log(ax) = \log a + \log x \quad \text{so} \quad k = \log a.$$

17. If profits are $r(t) = 0.5P(t)$ instead of $P(t)$, then profits are half the dollar level expected. If profits are $s(t) = P(0.5t)$ instead of $P(t)$, then profits are accruing half as fast as the projected rate.

Solutions for Section 5.5

1. First, substitute the y-intercept $(0, 5)$ into the standard form of the quadratic function to obtain:

$$5 = a(0)^2 + b(0) + c.$$

This yields $c = 5$. Next because of the symmetry of the parabola, $(-6, 5)$ is also on the graph. See Figure 5.40.

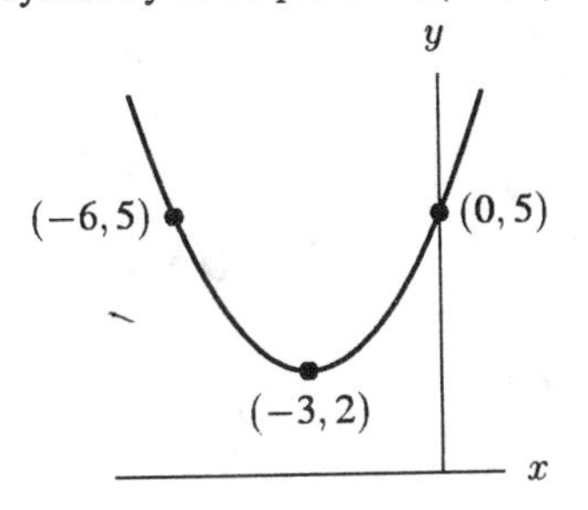

Figure 5.40

The vertex $(-3, 2)$ and the point $(-6, 5)$ when substituted into $y = ax^2 + bx + 5$ give the equations:

$$2 = 9a - 3b + 5$$
$$5 = 36a - 6b + 5.$$

Solving these two linear equations simultaneously yields $a = 1/3$ and $b = 2$. Therefore

$$y = m(x) = \frac{1}{3}x^2 + 2x + 5.$$

5. (a) See Figure 5.41.

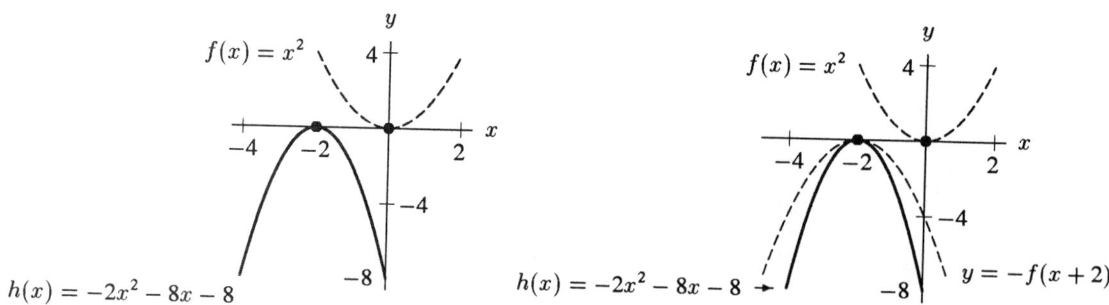

***Figure 5.41**:* The graphs of $f(x) = x^2$ and $h(x) = -2x^2 - 8x - 8$.

***Figure 5.42**:* The graph of $y = -f(x+2)$ is a compressed version of the graph of h. The graph of $y = -2f(x+2)$ is the same as $y = h(x)$

(b) From Figure 5.41, it appears as though the graph of h might be found by flipping the graph of $f(x) = x^2$ over the x-axis and then shifting it to the left. In other words, we might guess that the graph of h is given by $y = -f(x+2)$. This graph is shown in Figure 5.42. Notice, though, that $y = -f(x+2)$ is not as steep as the graph of $y = h(x)$. However, we can look at the y–intercepts of both graphs and notice that -4 must be stretched to -8. Thus, we can try applying a stretch factor of 2 to our guess. The resulting graph, given by

$$y = -2f(x+2),$$

does indeed match h.

We can verify the last result algebraically. If the graph of h is given by $y = -2f(x+2)$, then

$$\begin{aligned} y &= -2f(x+2) \\ &= -2(x+2)^2 \qquad \text{(because } f(x) = x^2\text{)} \\ &= -2(x^2 + 4x + 4) \\ &= -2x^2 - 8x - 8. \end{aligned}$$

This is the formula given for h.

9. The function appears quadratic with vertex at $(2, 0)$, so it could be of the form $y = a(x-2)^2$. For $x = 0$, $y = -4$, so $y = a(0-2)^2 = 4a = -4$ and $a = -1$. Thus $y = -(x-2)^2$ is a possible formula.

13. The graph of $y = x^2 - 10x + 25$ appears to be the graph of $y = x^2$ moved to the right by 5 units. See Figure 5.43. If this were so, then its formula would be $y = (x-5)^2$. Since $(x-5)^2 = x^2 - 10x + 25$, $y = x^2 - 10x + 25$ is, indeed, a horizontal shift of $y = x^2$.

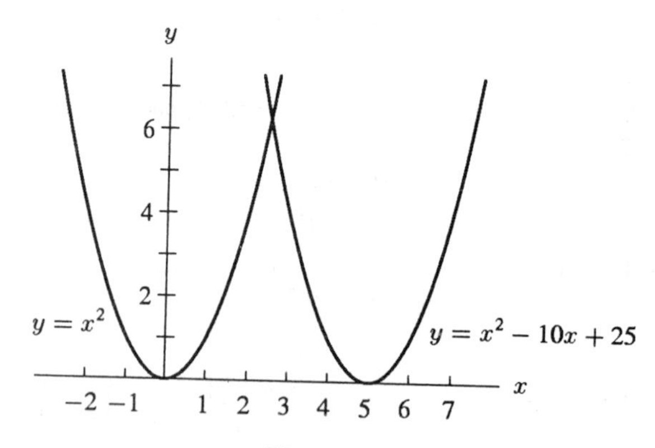

Figure 5.43

17. Factoring gives $y = -4cx + x^2 + 4c^2 = x^2 - 4ck + 4c^2 = (x - 2c)^2$. Since $c > 0$, this is the graph of $y = x^2$ shofted to the right $2c$ units. See Figure 5.44.

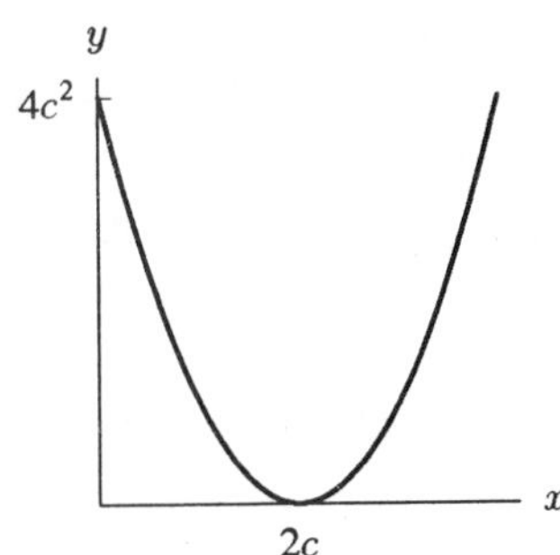

Figure 5.44: $y = -4cx + x^2 + 4c^2$ for $c > 0$

21. We can complete the square by taking $\frac{1}{2}$ of the coefficient of x, or $\frac{1}{2}(-12)$ and squaring the result. This gives $(\frac{1}{2} \cdot -12)^2 = (-6)^2 = 36$. We now use the number 36 to rewrite our formula for $r(x)$. We have

$$\begin{aligned} r(x) &= \underbrace{x^2 - 12x + (-6)^2}_{\text{completing the square}} - \underbrace{36}_{\text{compensating term}} + 28 \\ &= (x-6)^2 - 8. \end{aligned}$$

Thus, the vertex of this parabola is $(6, -8)$, and its axis of symmetry is $x = 6$.

25. Factoring out negative one (to make the coefficient of x^2 equal 1) and completing the square gives

$$\begin{aligned} y &= -1 \cdot \left(x^2 - 7x + \left(-\frac{7}{2}\right)^2 - \left(-\frac{7}{2}\right)^2 + 13\right) \\ &= -\left(x - \frac{7}{2}\right)^2 + \frac{49}{4} - 13 \\ &= -\left(x - \frac{7}{2}\right)^2 - \frac{3}{4}. \end{aligned}$$

Thus, the graph of this function is a downward-opening parabola with a vertex below the x-axis. Since the graph is below the x-axis and opens down, it does not intersect the x-axis. We conclude that this function has no zeros which are real numbers.

To see this algebraically, notice that the equation $y = 0$ has no real-valued solution, because solving

$$-\left(x - \frac{7}{2}\right)^2 - \frac{3}{4} = 0$$

gives

$$x = \frac{7}{2} \pm \sqrt{-\frac{3}{4}}$$

and $\sqrt{-\frac{3}{4}}$ is not a real number.

29. We will reverse Gwendolyn's actions. First, we can shift the parabola back two units to the right by replacing x in $y = (x-1)^2 + 3$ with $(x - 2)$. This gives

$$\begin{aligned} y &= ((x-2) - 1)^2 + 3 \\ &= (x-3)^2 + 3. \end{aligned}$$

We subtract 3 from this function to move the parabola down three units, so

$$\begin{aligned} y &= (x-3)^2 + 3 - 3 \\ &= (x-3)^2. \end{aligned}$$

Finally, to flip the parabola back across the horizontal axis, we multiply the function by -1. Thus, Gwendolyn's original equation was

$$y = -(x-3)^2.$$

33. (a) Since the function is quadratic, we take $y = ax^2 + bx + c$. We know that $y = 1$ when $x = 0$, so $y = a(0)^2 + b(0) + c = 1$. Thus, $c = 1$ and the formula is $y = ax^2 + bx + 1$. We know that $y = 3.01$ when $x = 1$, which gives us $y = a(1)^2 + b(1) + 1 = 3.01$, so $a + b + 1 = 3.01$, or $b = 2.01 - a$. Similarly, $y = 5.04$ when $x = 2$ suggests that $y = a(2)^2 + b(2) + 1 = 5.04$, which simplifies to $4a + 2b = 4.04$, or $2a + b = 2.02$. In this case, $b = 2.02 - 2a$. Since $b = 2.01 - a$ and $b = 2.02 - 2a$, we can say that

$$2.01 - a = 2.02 - 2a$$
$$a = 0.01$$

and

$$b = 2.01 - a = 2.01 - 0.01 = 2.$$

Since $a = 0.01, b = 2$, and $c = 1$, we know that

$$y = 0.01x^2 + 2x + 1$$

is the quadratic function passing through the first three data points. If $x = 50$, then

$$y = 0.01(50)^2 + 2(50) + 1 = 126,$$

so the fifth data point also satisfies the quadratic model.

(b) A linear function through (1, 3.01) and (2, 5.04) has slope $m = \frac{5.04-3.01}{2-1} = \frac{2.03}{1} = 2.03$, so $y = 2.03x + b$. We combine this with the knowledge that $(1, 3.01)$ lies on the line to get

$$3.01 = 2.03(1) + b$$
$$3.01 = 2.03 + b$$

so

$$b = 0.98.$$

Thus, a linear model using just the second two data points is

$$y = 2.03x + 0.98.$$

(c) Using this linear function, when $x = 3$,

$$y = 2.03(3) + 0.98 = 7.07.$$

Since the value of the quadratic function at $x = 3$ is 7.09, the difference between the quadratic and linear models at $x = 3$ is

$$7.09 - 7.07 = 0.02.$$

(d) At $x = 50$, the linear function has a value of $y = 2.03(50) + 0.98 = 102.48$. The quadratic function gives 126 when $x = 50$. The difference in output is $126 - 102.48 = 23.52$.

(e) If we want the difference to be less than 0.05, we want

$$|(0.01x^2 + 2x + 1) - (2.03x + 0.98)| = |0.01x^2 - 0.03x + 0.02| \leq 0.05.$$

Using a graphing calculator or computer, we graph $y = 0.01x^2 - 0.03x + 0.02$ and look for the values of x for which $|y| \leq 0.05$. These occur when x is between -0.791 and 3.791.

37.

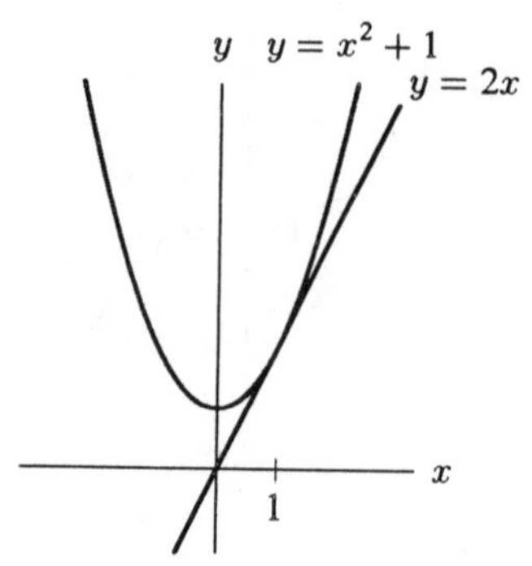

Figure 5.45: Graphs of $y = 2x$ and $y = x^2 + 1$

From the graph in Figure 5.45, it appears that for any x, the graph of $y = 2x$ is below the graph of $y = x^2 + 1$, except at $x = 1$ where they are possibly equal. This suggests that $x^2 + 1$ is greater than $2x$, except possibly at $x = 1$, or in terms of inequalities,

$$x^2 + 1 \geq 2x.$$

Subtracting $2x$ from both sides gives

$$x^2 - 2x + 1 \geq 0.$$

Factoring gives

$$(x - 1)^2 \geq 0.$$

This inequality is always true. Since the last statement is equivalent to saying $x^2 + 1 \geq 2x$, we have shown why our conjecture is true.

41. (a) Factoring gives $h(t) = -16t^2 + 16Tt = 16t(T - t)$. Since $h(t) \geq 0$ only for $0 \leq t \leq T$, the model makes sense only for these values of t.
(b) The times $t = 0$ and $t = T$ give the start and end of the jump. The maximum height occurs halfway in between, at $t = T/2$.
(c) Since $h(t) = 16t(T - t)$, we have

$$h\left(\frac{T}{2}\right) = 16\left(\frac{T}{2}\right)\left(T - \frac{T}{2}\right) = 4T^2.$$

Solutions for Chapter 5 Review

1. (a) The input is $2x = 2 \cdot 2 = 4$.
(b) The input is $\frac{1}{2}x = \frac{1}{2} \cdot 2 = 1$.
(c) The input is $x + 3 = 2 + 3 = 5$.
(d) The input is $-x = -2$.

5. (a) They are different functions. With $e^x + 1$, the calculator would evaluate e^x for some value of x and then add one to the result. For e^{x+1}, the calculator first adds 1 to the chosen value of x, then evaluates e raised to that sum. Graphically, $y = e^x + 1$ is the graph of $y = e^x$ shifted up by 1, while $y = e^{x+1}$ is the graph of $y = e^x$ shifted to the left by 1.
(b) The calculator would assume $e^x + 1$, since it follows order of operations. It therefore exponentiates first, then adds. To get the function e^{x+1} one should enter e^ $(x + 1)$.

9.

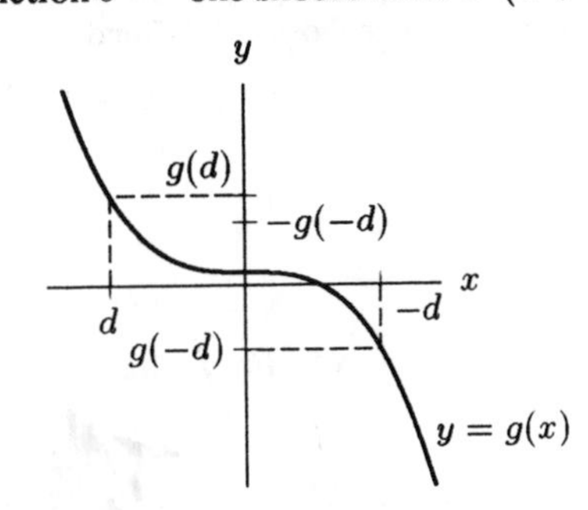

Figure 5.46

13. The graphs are shown in Figure 5.47:

(a)

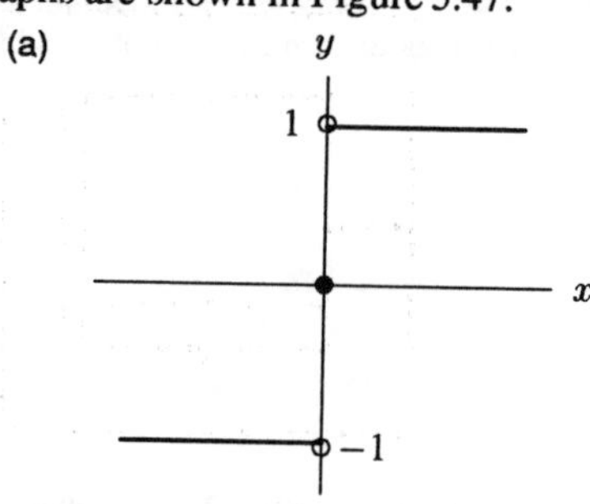

(b)

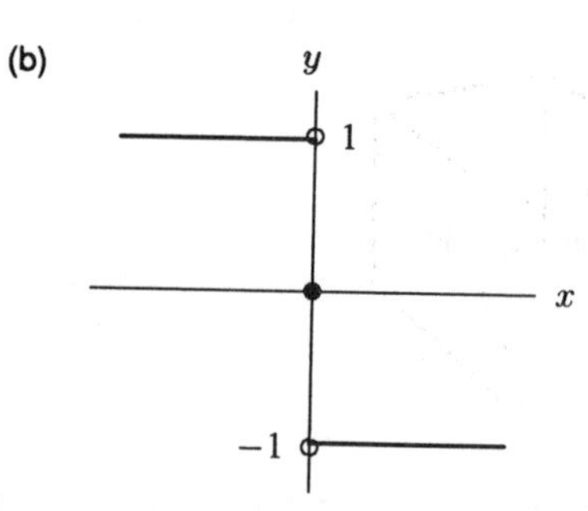

(c)

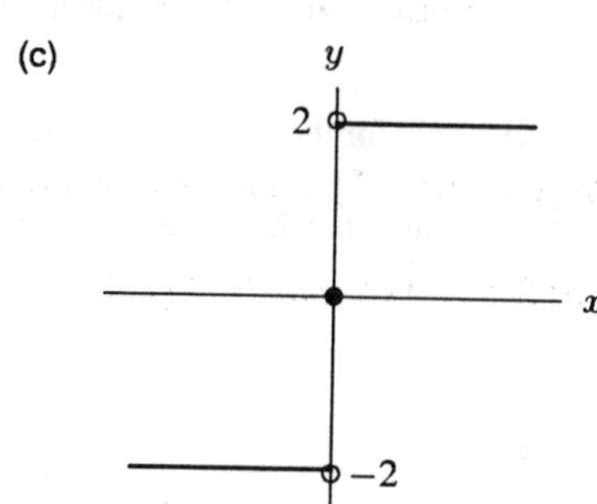

(d)

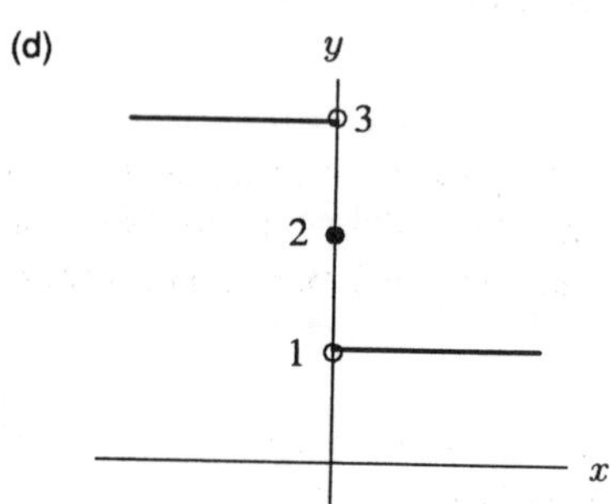

Figure 5.47

17. By the quadratic formula, $\left(\dfrac{-b+\sqrt{b^2-4ac}}{2a}\right)$ is a root of ax^2+bx+c. Thus

$$f\left(\frac{-b+\sqrt{b^2-4ac}}{2a}\right)=0.$$

21. Assuming f is linear between 10 and 20, we get

$$\frac{6400-f(10)}{p-10}=\frac{f(20)-f(10)}{20-10},$$

$$\text{or}\quad \frac{400}{p-10}=\frac{800}{10}.$$

Solving for p yields $p=15$. Assuming f is linear between 20 and 30, we get

$$\frac{q-f(20)}{26-20}=\frac{f(30)-f(20)}{30-20},$$

$$\text{or}\quad \frac{q-6800}{6}=\frac{650}{10}.$$

Solving for q yields $q=7190$.

25. (a) There is a vertical stretch of 3 so

$$y=3h(x).$$

(b) We have a reflection through the x-axis and a horizontal shift to the right by 1.

$$y=-h(x-1)$$

(c) There is a reflection through the y-axis, a horizontal compression by a factor of 2, a horizontal shift to the right by 1 unit, and a reflection through the x-axis. Combining these transformations we get

$$y=-h(-2(x-1))\quad \text{or}\quad y=-h(2-2x).$$

29.

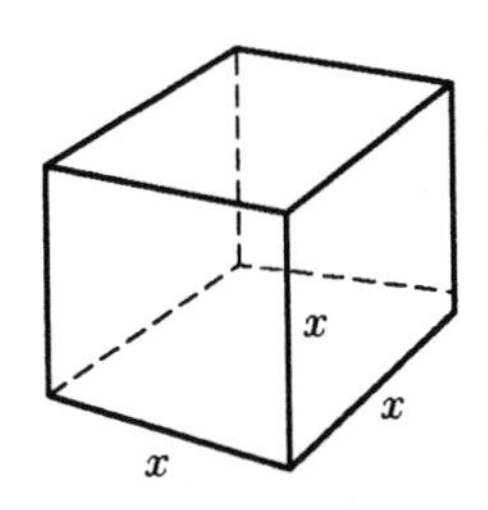

Figure 5.48

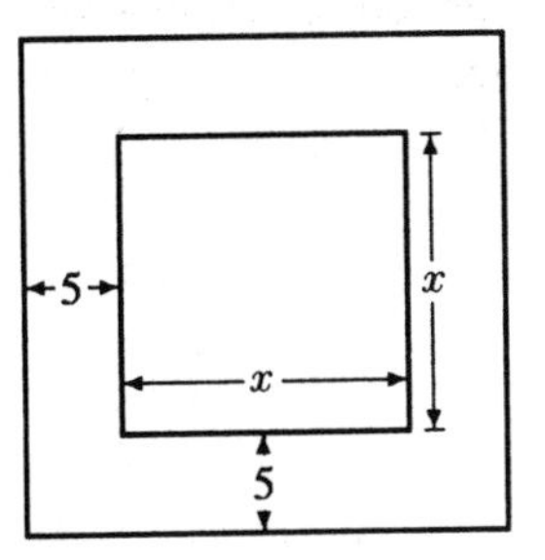

Figure 5.49: cross sectioned view

(a) $L(x) = 12x$
(b) $L(x) - 6$ cm of tape would be short by 6 cm of doing the job. $L(x - 6)$ cm of tape would be short by 6 cm for each edge (a total of 72 cm short all together).
(c) Each edge would require 2 extra cm of tape, so the required amount would be $L(x + 2)$.
(d) $S(x) = 6x^2$.
(e) $V(x) = x^3$.
(f) Surface area of larger box $= S(x + 10)$.
(g) Volume of larger box $= V(x + 10)$.
(h) Tape on edge of larger box $= L(x + 10)$.
(i) Double taping the outer box would require $2L(x + 10)$, or $L(2(x + 10))$. ($L(2x)$ will give the same function values as $2L(x)$.)
(j) A box with edge length 20% longer than x has length $x + 0.20x = 1.2x$; the "taping function" should input that edge length, hence $L(1.2x)$ does the job. (But note that $1.2L(x)$ will also work in this case.)

CHAPTER SIX

Solutions for Section 6.1

1. The wheel will complete two full revolutions after 20 minutes, so the function is graphed on the interval $0 \le t \le 20$. See Figure 6.1.

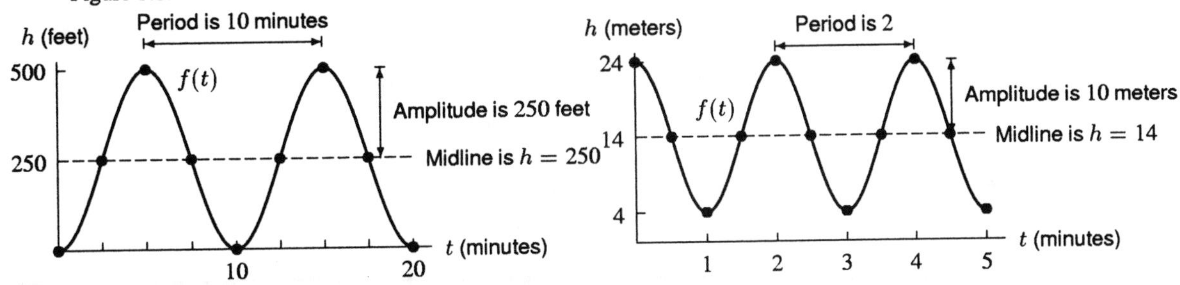

Figure 6.1: Graph of $h = f(t), 0 \le t \le 20$

Figure 6.2: Graph of $h = f(t), 0 \le t \le 5$

5. See Figure 6.2.

9. At $t = 0$, we see $h = 20$, so you are level with the center of the wheel. Your initial position is at three o'clock (or nine o'clock) and initially you are rising. On the interval $0 \le t \le 7$ the wheel completes seven fourths of a revolution. Therefore, if p is the period, we know that
$$\frac{7}{4}p = 7$$
which gives $p = 4$. This means that the ferris wheel takes 4 minutes to complete one full revolution. The minimum value of the function is $h = 5$, which means that you get on and get off of the wheel from a 5 meter platform. The maximum height above the midline is 15 meters, so the wheel's diameter is 30 meters. Notice that the wheel completes a total 2.75 cycles. Since each period is 4 minutes long, you ride the wheel for $4(2.75) = 11$ minutes.

13. The midline of f is $d = 10$. The period of f is 1, the amplitude 4 cm, and its minimum and maximum values are 6 cm and 14 cm, respectively. The fact that $f(t)$ is wave-shaped means that the spring is bobbing up and down, or *oscillating*. The fact that the period of f is 1 means that it takes the weight one second to complete one oscillation and return to its original position. Studying the graph, we see that it takes the weight 0.25 seconds to move from its initial position at the midline to its maximum at $d = 14$, where it is farthest from the ceiling (and the spring is at its maximum extension). It takes another 0.25 seconds to return to its initial position at $d = 10$ cm. It takes another 0.25 seconds to rise up to its closest distance from the ceiling at $d = 6$ (the minimum extension of the spring). In 0.25 seconds more it moves back down to its initial position at $d = 10$. (This sequence of motions by the weight, completed in one second, represents one full oscillation.) Since Figure 6.10 of the text gives 3 full periods of $f(t)$, it represents the 3 complete oscillations made by the weight in 3 seconds.

17. (a) Two possible answers are shown in Figures 6.3 and 6.4.

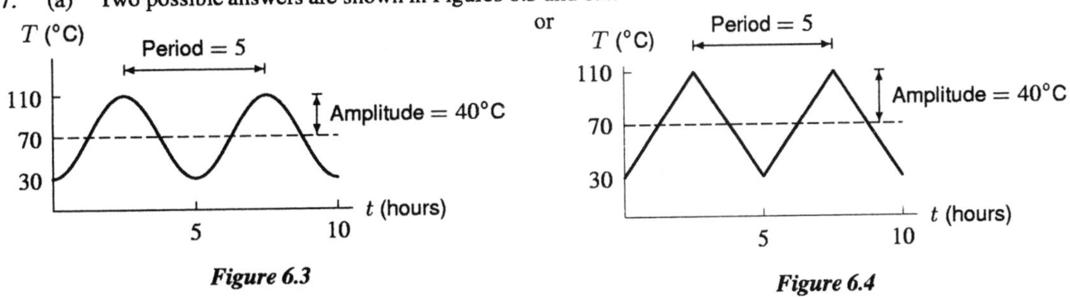

Figure 6.3

Figure 6.4

(b) The period is 5 hours. This is the time required for the temperature to cycle from 30° to 110° and back to 30°. The midline, or average temperature, is $T = (110 + 30)/2 = 70°$. The amplitude is 40° since this is the amount of temperature variation (up or down) from the average.

21. (a) See Figure 6.5. Notice that the function is only *approximately* periodic.

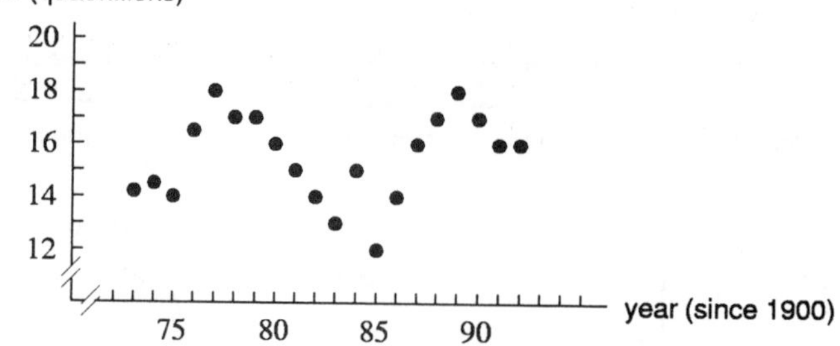

Figure 6.5: U.S. Imports of Petroleum

(b) There appears to be a peak in imports every twelve years.
(c) Midline: $y = 15$; Amplitude: 3; Period: 12

Solutions for Section 6.2

1.

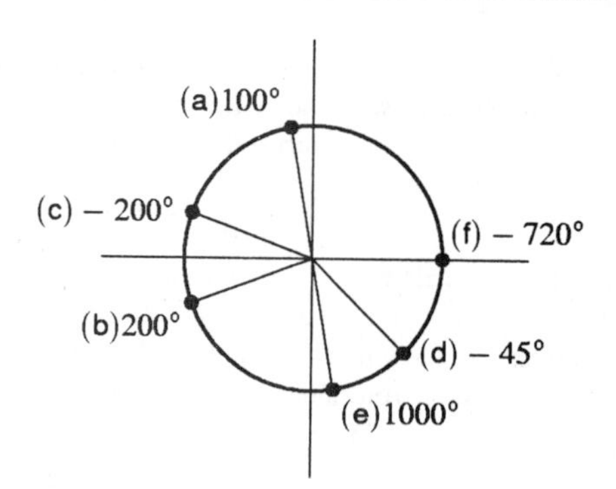

Figure 6.6

(a) $(\cos 100°, \sin 100°) = (-0.174, 0.985)$
(b) $(\cos 200°, \sin 200°) = (-0.940, -0.342)$
(c) $(\cos(-200°), \sin(-200°)) = (-0.940, 0.342)$
(d) $(\cos(-45°), \sin(-45°)) = (0.707, -0.707)$
(e) $(\cos 1000°, \sin 1000°) = (0.174, -0.985)$
(f) $(\cos 720°, \sin 720°) = (1, 0)$

5. To locate the points P, Q, and R, we mark off their respective angles, $540°$, $-180°$, and $450°$, by measuring these angles from the positive x-axis in the counterclockwise direction if the angle is positive and in the clockwise direction if the angle is negative. See Figure 6.7.

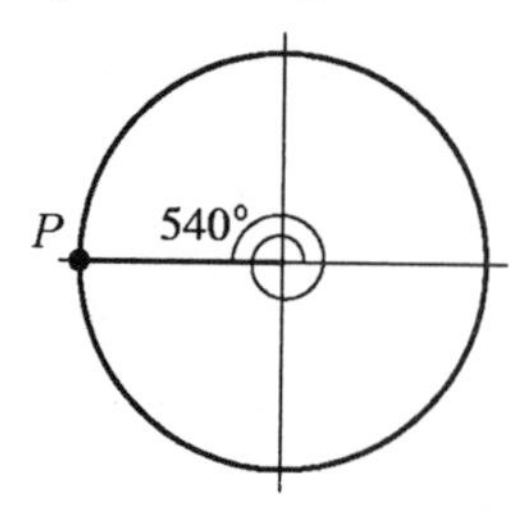

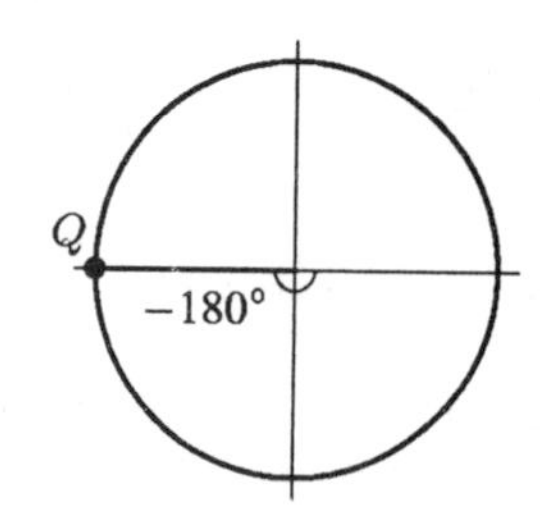

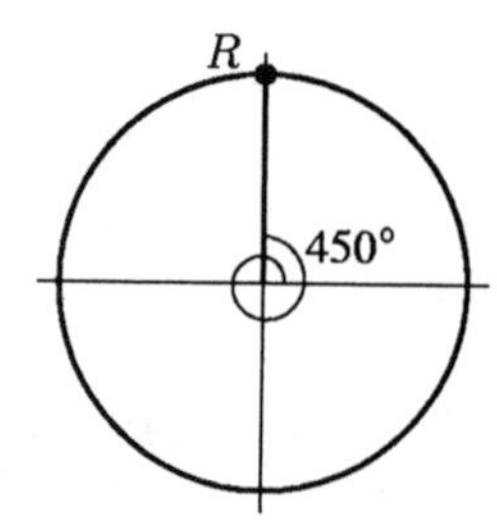

Figure 6.7

$P = (-1, 0)$, $Q = (-1, 0)$, $R = (0, 1)$

9. (a) As we see from Figure 6.8, the angle 135° specifies a point P' on the unit circle directly across the y-axis from the point P. Thus, P' has the same y-coordinate as P, but its x-coordinate is opposite in sign to the x-coordinate of P. Therefore, $\sin 135° = 0.71$, and $\cos 135° = -0.71$.
   (b) As we see from Figure 6.9, the angle 285° specifies a point Q' on the unit circle directly across the x-axis from the point Q. Thus, Q' has the same x-coordinate as Q, but its y-coordinate is opposite in sign to the y-coordinate of Q. Therefore, $\sin 285° = -0.97$, and $\cos 285° = 0.26$.

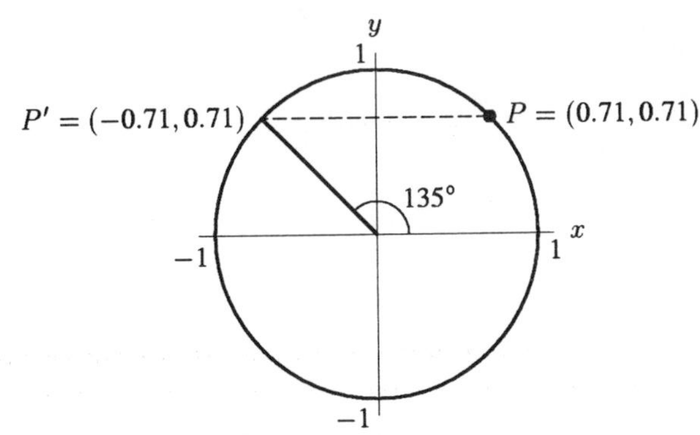

Figure 6.8: The sine and cosine of 135° can be found by referring to the sine and cosine of 45°

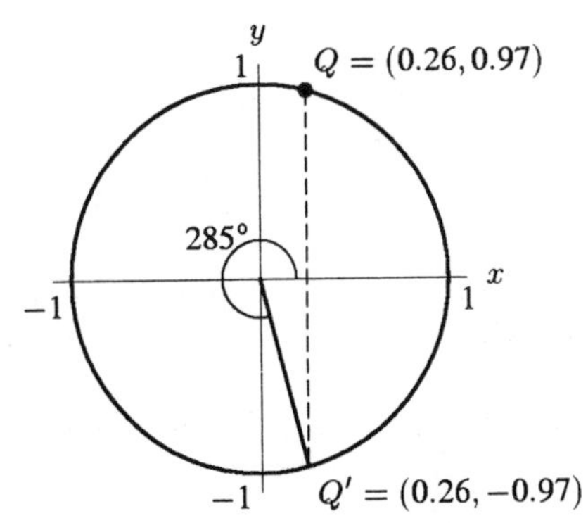

Figure 6.9: The sine and cosine of 285° can be found by referring to the sine and cosine of 75°

13. We know that the four panels are evenly spread out, so the angle between two neighboring panels must be 90°. We also know that each panel has the same length. Thus, the triangle going through points B, C and the origin, O, is a $45° - 45° - 90°$ triangle whose hypotenuse is the distance from B to C. We know that the length of each of the other sides is 2 meters. So using the Pythagorean theorem, we have

$$\text{Distance from } B \text{ to } C = \sqrt{2^2 + 2^2} = 2\sqrt{2} \approx 2.83 \text{ meters.}$$

17. (a) $\sin(\theta + 360°) = \sin\theta = a$, since the sine function is periodic with a period of 360°.
   (b) $\sin(\theta + 180°) = -a$. (A point on the unit circle given by the angle $\theta + 180°$ diametrically opposite the point given by the angle θ. So the y-coordinates of these two points are opposite in sign, but equal in magnitude.)
   (c) $\cos(90° - \theta) = \sin\theta = a$. This is most easily seen from the right triangles in Figure 6.10.

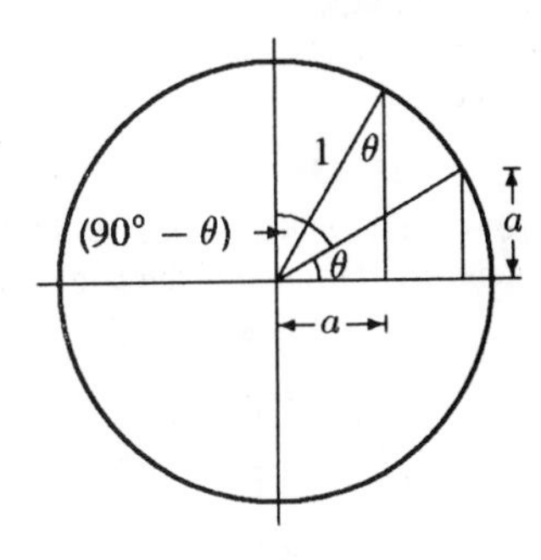

Figure 6.10

   (d) $\sin(180° - \theta) = a$. (A point on the unit circle given by the angle $180° - \theta$ has a y-coordinate equal to the y-coordinate of the point on the unit circle given by θ.)
   (e) $\sin(360° - \theta) = -a$. (A point on the unit circle given the the angle $360° - \theta$ has a y-coordinate of the same magnitude as the y-coordinate of the point on the unit circle given by θ, but is of opposite sign.)
   (f) $\cos(270° - \theta) = -\sin\theta = -a$.

21. See Figure 6.11. Since the diameter is 120 mm, the radius is 60 mm. The coordinates of the outer edge point, A, on the x-axis is $(60, 0)$. Similarly the inner edge at point B has coordinates $(7.5, 0)$.

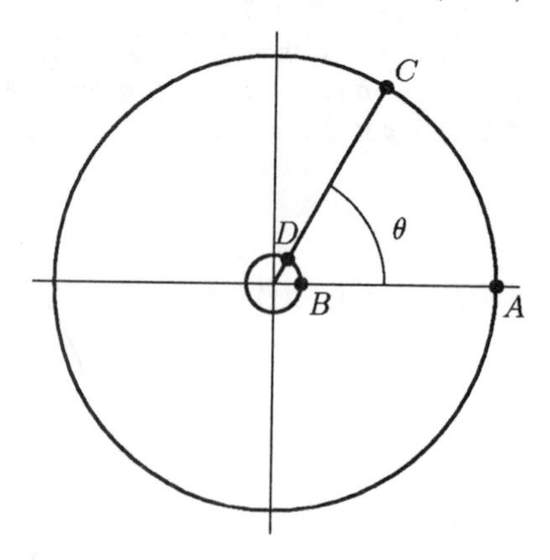

Figure 6.11

Points C and D are at an angle θ from the x-axis and have coordinates of the form $(r\cos\theta, r\sin\theta)$. For the outer edge, $r = 60$ so $C = (60\cos\theta, 60\sin\theta)$. The inner edge has $r = 7.5$, so $D = (7.5\cos\theta, 7.5\sin\theta)$.

Solutions for Section 6.3

1. To convert $45°$ to radians, multiply by $\pi/180°$:

$$45°\left(\frac{\pi}{180°}\right) = \left(\frac{45°}{180°}\right)\pi = \frac{\pi}{4}.$$

Thus we say that the radian measure of a $45°$ angle is $\pi/4$.

5. (a) $30 \cdot \frac{\pi}{180} = \frac{\pi}{6}$ or 0.52
   (b) $120 \cdot \frac{\pi}{180} = \frac{2\pi}{3}$ or 2.09
   (c) $200 \cdot \frac{\pi}{180} = \frac{10\pi}{9}$ or 3.49
   (d) $315 \cdot \frac{\pi}{180} = \frac{7\pi}{4}$ or 5.50

9. (a) $-2\pi/3 < 2/3 < 2\pi/3 < 2.3$
   (b) $\cos 2.3 < \cos(-2\pi/3) = \cos(2\pi/3) < \cos(2/3)$

13. First $225°$ has to be converted to radian measure:

$$225 \cdot \frac{\pi}{180} = \frac{5\pi}{4}.$$

Using $s = r\theta$ gives

$$s = 4 \cdot \frac{5\pi}{4} = 5\pi \text{ feet.}$$

17. Since the ant traveled three units on the unit circle, the traversed arc must be spanned by an angle of three radians. Thus the ant's coordinates must be

$$(\cos 3, \sin 3) \approx (-0.99, 0.14).$$

21. (a) Using the formula $s = \theta r$ with $r = 38/2 = 19$ cm we find

$$s = \text{ Arc length } = (19)(3.83) = 72.77 \text{ cm.}$$

   (b) Using $s = \varphi r$ with s and r known, we have $3.83 = \varphi(19)$. Thus $\varphi = 3.83/19 \approx 0.2$ radians.

25. Using $s = r\theta$, we know the arc length $s = 600$ and $r = 3960 + 500$. Therefore $\theta = 600/4460 \approx 0.1345$ radians.

29. Make a table using your calculator to see that $\cos t$ is decreasing and the values of t are increasing.

t	0	0.1	0.2	0.3	0.4	0.5	0.6	0.7	0.8	0.9
$\cos t$	1	0.9950	0.9801	0.9533	0.9211	0.8776	0.8253	0.7648	0.6967	0.6216

Use a more refined table to see $t \approx 0.74$. (See Table 6.1.)

TABLE 6.1

t	0.70	0.71	0.72	0.73	0.74	0.75	0.76
$\cos t$	0.7648	0.7584	0.7518	0.7451	0.7385	0.7317	0.7248

Alternatively, consider the graphs of $y = t$ and $y = \cos t$ in Figure 6.12. They intersect at a point in the first quadrant, so for the t-coordinate of this point, $t = \cos t$. Trace with a calculator to find $t \approx 0.74$.

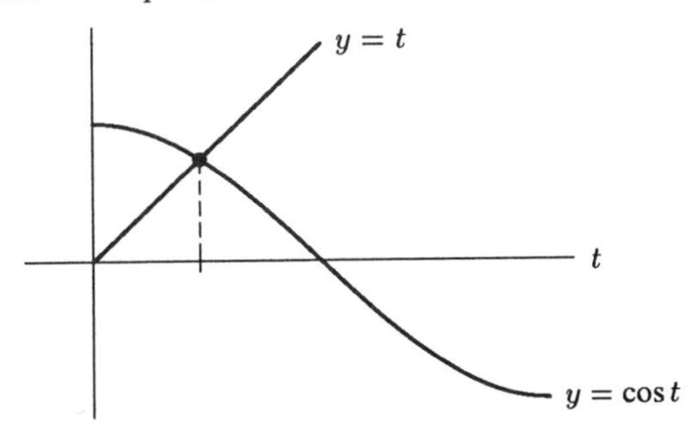

Figure 6.12

Solutions for Section 6.4

1.

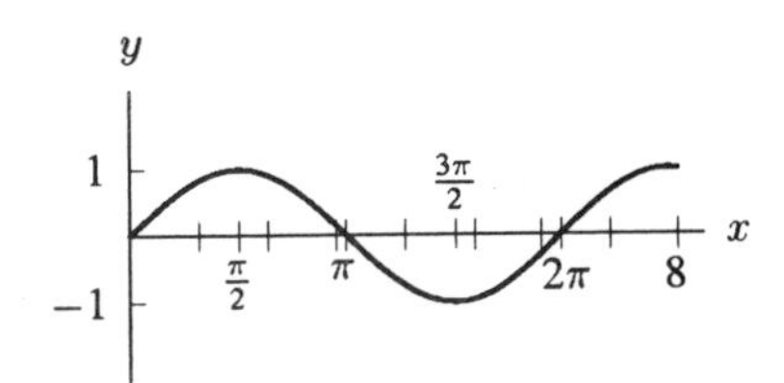

Figure 6.13

5. $g(x) = 2\sin x$, $a = \pi$ and $b = 2$.

9. We can sketch these graphs using a calculator or computer. Figure 6.14 gives a graph of $y = \sin\theta$, together with the graphs of $y = 0.5\sin\theta$ and $y = -2\sin\theta$, where θ is in radians and $0 \le \theta \le 2\pi$.

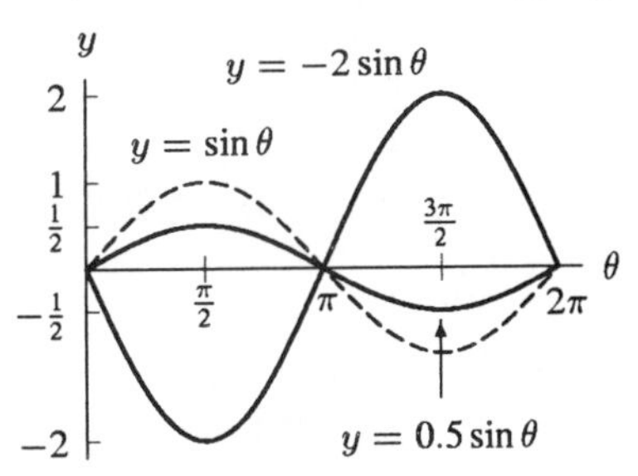

Figure 6.14

These graphs are similar but not the same. The amplitude of $y = 0.5\sin\theta$ is 0.5 and the amplitude of $y = -2\sin\theta$ is 2. The graph of $y = -2\sin\theta$ is vertically reflected relative to the other two graphs. These observations are consistent with the fact that the constant A in the equation

$$y = A\sin\theta$$

may result in a vertical stretching or shrinking and/or a reflection over the x-axis. Note that all three graphs have a period of 2π.

13. (a) (i) p
(ii) s
(iii) q
(iv) r

(b)

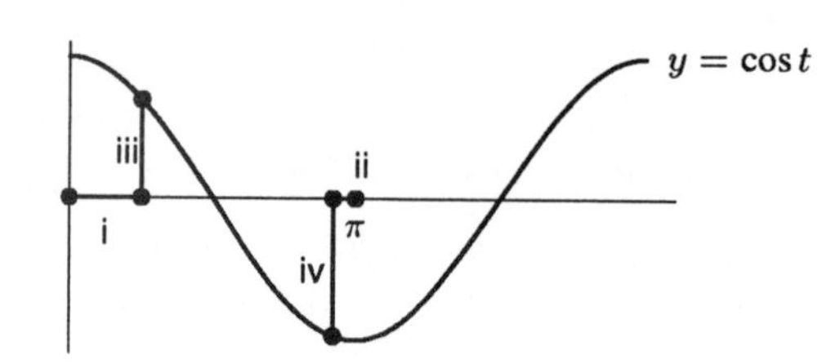

Figure 6.15

17. (a) (i)

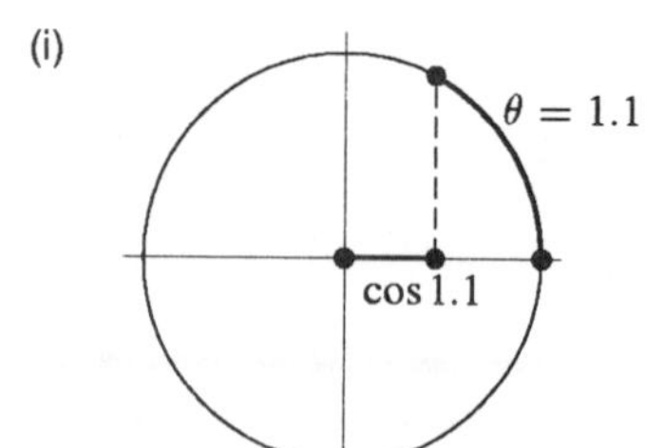

(ii)

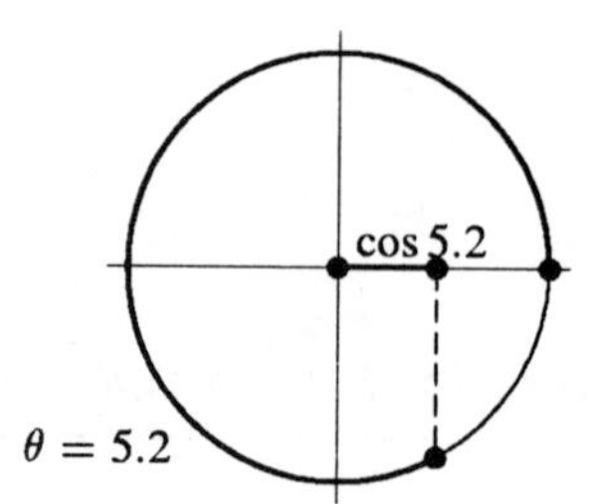

(b)

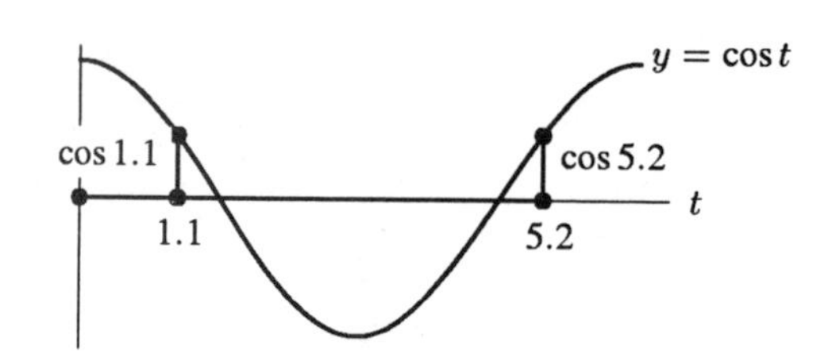

Figure 6.16

21. (a) Slope, m, of segment joining S and T:

$$m = \frac{\sin(a+h) - \sin a}{(a+h) - a} = \frac{\sin(a+h) - \sin a}{h}$$

(b) If $a = 1.7$ and $h = 0.05$,

$$m = \frac{\sin 1.75 - \sin 1.7}{1.75 - 1.7} \approx -0.15$$

Solutions for Section 6.5

1. (a) The function $y = \sin(-t)$ is periodic, and its period is 2π. The function begins repeating every 2π units, as is clear from its graph. Recall that $f(-x)$ is a reflection about the y-axis of the graph of $f(x)$, so the periods for $\sin(t)$ and $\sin(-t)$ are the same. See Figure 6.17.

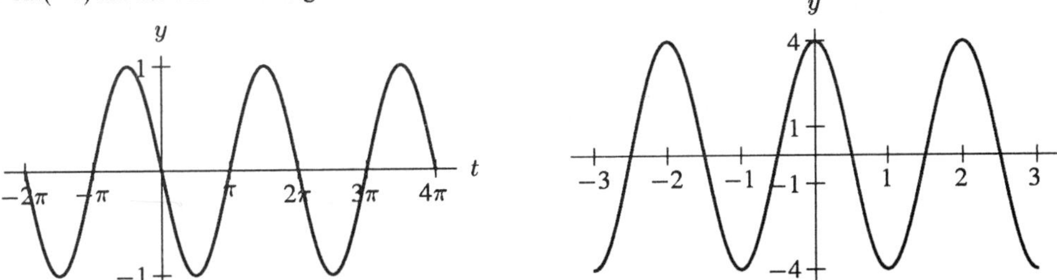

Figure 6.17: $y = \sin(-t)$ *Figure 6.18*: $y = 4\cos(\pi t)$

(b) The function $y = 4\cos(\pi t)$ is periodic, and its period is 2. This is because when $0 \leq t \leq 2$, we have $0 \leq \pi t \leq 2\pi$ and the cosine function has period 2π. Note the amplitude of $4\cos(\pi t)$ is 4, but changing the amplitude does not affect the period. See Figure 6.18.

(c) The function $y = \sin(t) + t$ is not periodic, because as t gets large, $\sin(t) + t$ gets large as well. In fact, since $\sin(t)$ varies from -1 to 1, y is always between $t - 1$ and $t + 1$. So the values of y cannot repeat. See Figure 6.19.

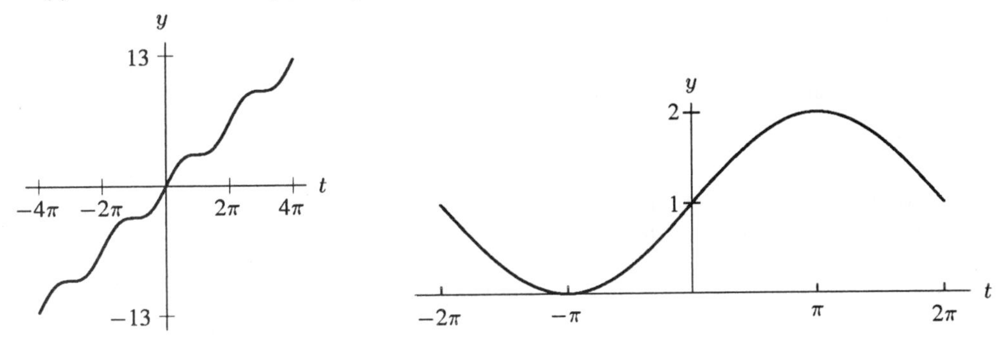

Figure 6.19: $y = \sin(t) + t$ *Figure 6.20*: $y = \sin(t/2) + 1$

(d) In general $f(x)$ and $f(x) + c$ will have the same period if they are periodic. The function $y = \sin(\frac{t}{2}) + 1$ is periodic, because $\sin(\frac{t}{2})$ is periodic. Since $\sin(t/2)$ completes one cycle for $0 \leq t/2 \leq 2\pi$, or $0 \leq t \leq 4\pi$, we see the period of $y = \sin(t/2) + 1$ is 4π. See Figure 6.20.

5.

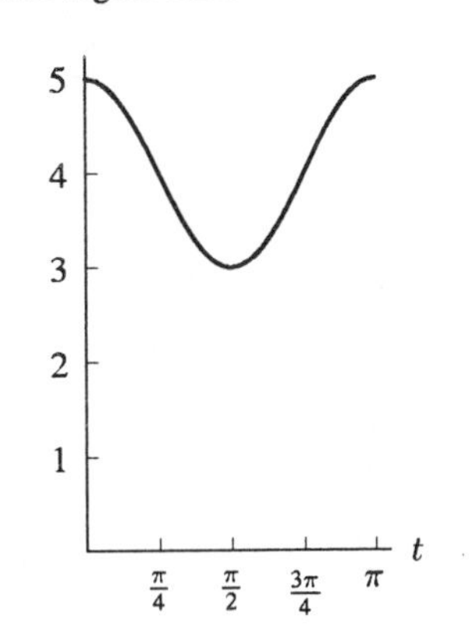

Figure 6.21: $y = \cos(2t) + 4$

9. The amplitude is 1, the period is $\frac{2\pi}{1/4} = 8\pi$, the phase shift is $\frac{\pi}{4}$, and

$$\text{Horizontal shift} = \frac{\pi/4}{1/4} = \pi.$$

Since the horizontal shift is positive, the graph of $y = \cos(t/4)$ is shifted π units to the right to give the graph in Figure 6.22. This is also seen from Figure 6.22 since maximum of the function is at $t = \pi$ instead of $t = 0$.

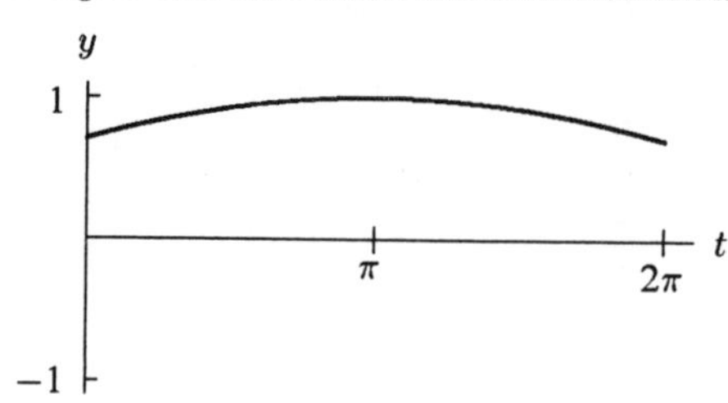

Figure 6.22: $y = \cos((t - \pi)/4)$

13. This function resembles an inverted cosine curve in that it attains its minimum value when $t = 0$. We know that the smallest value it attains is 0 and that its midline is $y = 2$. Thus its amplitude is 2 and it is shifted upward by two units. It has a period of 4π. Thus in the equation

$$g(t) = -A\cos(Bt) + D$$

we know that $A = -2$, $D = 2$, and

$$4\pi = \text{period} = \frac{2\pi}{B}.$$

So $B = 1/2$, and then

$$g(t) = -2\cos\left(\frac{t}{2}\right) + 2.$$

17. The graph is a horizontally stretched cosine function that is vertically shifted down 4 units, but is not horizontally shifted. The midline $y = -4$ is given. The amplitude is 1. The period is 13, so $B = 2\pi/13$. Thus

$$y = \cos\left(\frac{2\pi}{13}\theta\right) - 4.$$

21. Because the period of $\sin x$ is 2π, and the period of $\sin 2x$ is π, so from the figure in the problem we see that

$$f(x) = \sin x.$$

The points on the graph are $a = \pi/2$, $b = \pi$, $c = 3\pi/2$, $d = 2\pi$, and $e = 1$.

25. $f(t) = 14 + 10\sin\left(\pi t + \frac{\pi}{2}\right)$

29. $f(t) = 20 + 15\sin\left(\frac{\pi}{2}t + \frac{\pi}{2}\right)$

33. The graph of $y = (\sin x)^2$ is in Figure 6.23.

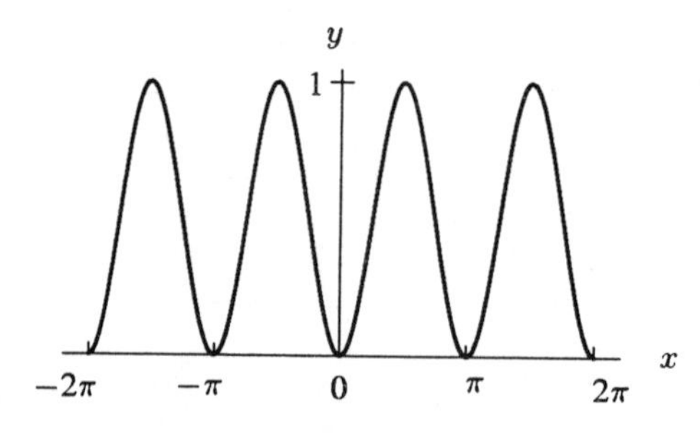

Figure 6.23

The maximum of this function is 1 and the minimum is 0. Thus

$$\text{Midline height} = k = \frac{\text{Max} + \text{Min}}{2} = \frac{1}{2}.$$

Then

$$\text{Amplitude} = |A| = \text{Max} - k = \frac{1}{2}.$$

The period of this function appears to be π. So

$$\pi = \text{Period} = \frac{2\pi}{B}.$$

Thus $B = 2$. Also, looking at the graph we see that it appears to be a vertically reflected cosine function since the minimum occurs at $x = 0$. Thus we have

$$(\sin x)^2 = -\frac{1}{2}\cos(2x) + \frac{1}{2}.$$

37. Both f and g have periods of 1, amplitudes of 1, and midlines $y = 0$.

41.

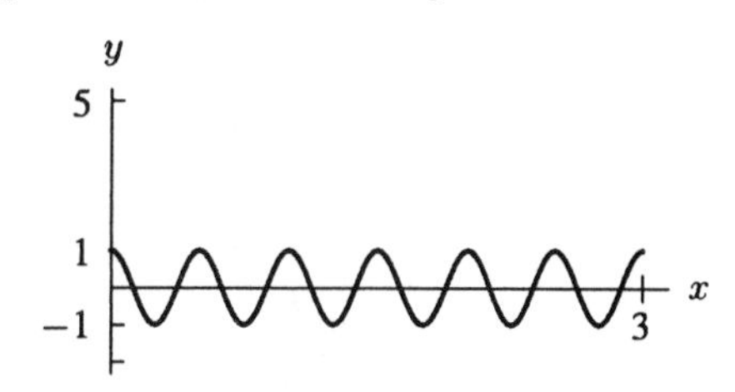

45. This function has an amplitude of 2 and a period of 3, and resembles vertically reflected cosine graph. Thus $y = -2g(x/3)$.

49. (a)

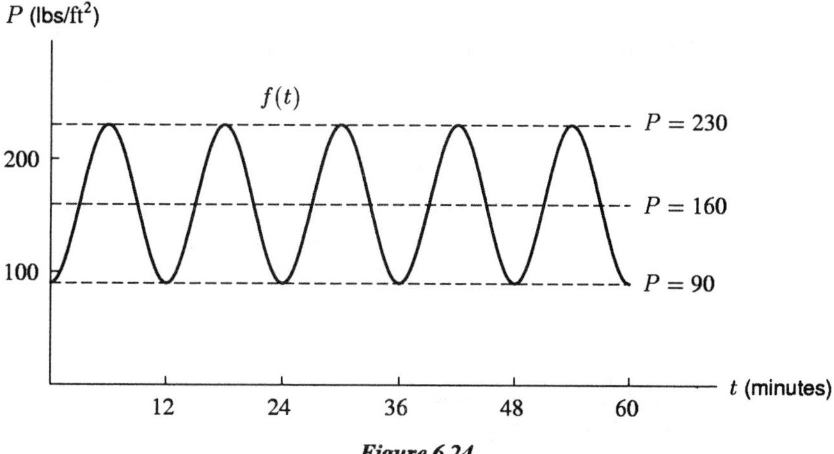

Figure 6.24

This function is a vertically reflected cosine function which has been vertically shifted. Thus the function for this equation will be of the form

$$P = f(t) = -A\cos(Bt) + k.$$

(b) The midline value is $k = (90 + 230)/2 = 160$.
The amplitude is $|A| = 230 - 160 = 70$.
A complete oscillation is made each 12 minutes, so the period is 12. This means $B = 2\pi/12 = \pi/6$. Thus $P = f(t) = -70\cos(\pi t/6) + 160$.

(c) Graphing $P = f(t)$ on a calculator for $0 \le t \le 2$ and $90 \le P \le 230$, we see that $P = f(t)$ first equals 115 when $t \approx 1.67$ minutes.

53. The petroleum import data is graphed in Figure 6.25.

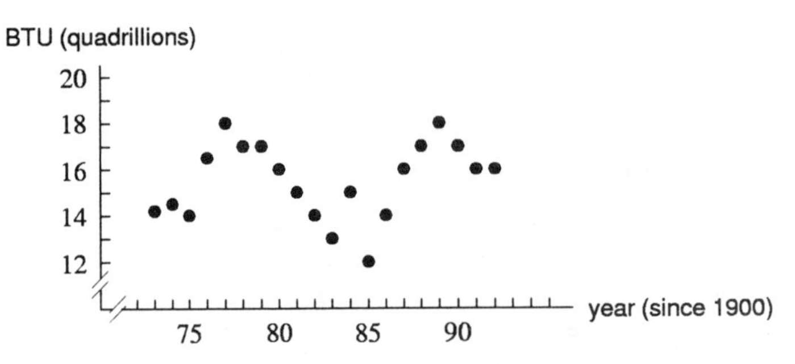

Figure 6.25: US Imports of Petroleum

We start with a sine function of the form

$$f(t) = A\sin(B(t-h)) + k.$$

Since the maximum here is 18 and the minimum is 12, the midline value is $k = (18+12)/2 = 15$. The amplitude is then $A = 18 - 15 = 3$. The period, measured peak to peak, is 12. So $B = 2\pi/12 = \pi/6$. Lastly, we calculate h, the horizontal shift to the right. Our data are close to the midline value for $t \approx 74$, whereas $\sin t$ is at its midline value for $t = 0$. So $h = 74$ and the equation is

$$f(t) = 3\sin\frac{\pi}{6}(t-74) + 15.$$

We can check our formula by graphing it and seeing how close it comes to the data points. See Figure 6.26.

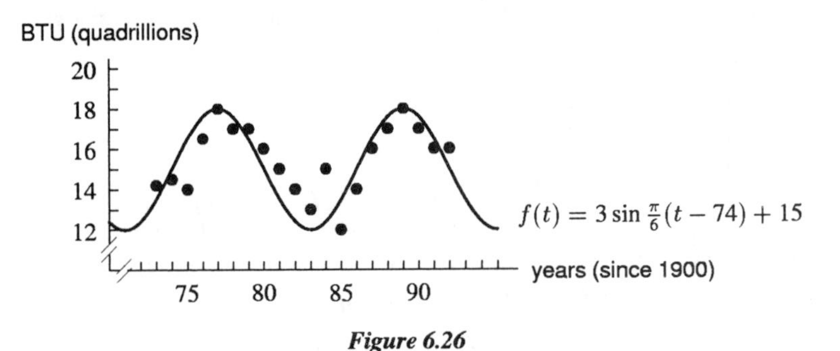

Figure 6.26

Solutions for Section 6.6

1. $\sin 0^\circ = 0$, $\cos 0^\circ = 1$, $\tan 0^\circ = \sin 0^\circ / \cos 0^\circ = 0/1 = 0$.

5. Since 135° is in the second quadrant,
$$\tan 135^\circ = -\tan 45^\circ = -1.$$

9. $\sqrt{3}$

13. Since $13\pi/6$ is in the first quadrant,
$$\tan\left(\frac{13\pi}{6}\right) = \tan\frac{\pi}{6} = \frac{1}{\sqrt{3}}.$$

17. Since $y = \sin\theta$, we can construct the following triangle:

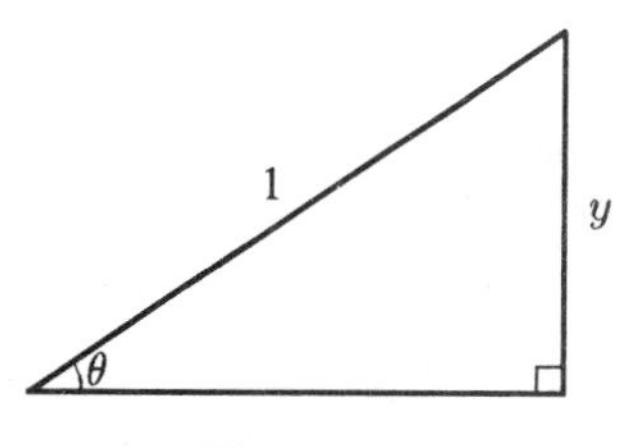

Figure 6.27

The adjacent side, using the Pythagorean theorem, has length $\sqrt{1-y^2}$. So, $\cos\theta = \frac{\text{adj}}{\text{hyp}} = \frac{\sqrt{1-y^2}}{1} = \sqrt{1-y^2}$.

21. First notice that $\tan\theta = \frac{x}{9}$ so $\tan\theta = \sin\theta/\cos\theta = x/9$, so $\sin\theta = x/9 \cdot \cos\theta$. Now to find $\cos\theta$ by using $1 = \sin^2\theta + \cos^2\theta = (x^2/81)\cos^2\theta + \cos^2\theta = \cos^2\theta(x^2/81+1)$, so $\cos^2\theta = 81/(x^2+81)$ and $\cos\theta = 9/\sqrt{x^2+81}$. Thus, $\sin\theta = (x/9)\cdot(9/\sqrt{x^2+81}) = x/\sqrt{x^2+81}$.

25. (a) The graph of $y = \tan t$ has vertical asymptotes at odd multiples of $\pi/2$, that is, at $\pi/2, 3\pi/2, 5\pi/2$, etc., and their negatives. The graph of $y = \cos t$ has t-intercepts at the same values.
    (b) The graph of $y = \tan t$ has t-intercepts at multiples of π, that is, at $0, \pm\pi, \pm 2\pi, \pm 3\pi$, etc. The graph of $y = \sin t$ has t-intercepts at the same values.

29. The angle spanned by the arc shown is $\theta = s/r = 10/5 = 2$ radians, so $v = r\sin\theta = 5\sin 2$. Since u is positive as it is a length, $u = -r\cos\theta = -5\cos 2$, because $\cos 2$ is negative. By the Pythagorean theorem,

$$\begin{aligned} w^2 &= v^2 + (5+u)^2 \\ &= v^2 + u^2 + 10u + 25 \\ &= 25\sin^2 2 + 25\cos^2 2 + 10u + 25 \\ &= 50 + 10u, \end{aligned}$$

and

$$w = \sqrt{50+10u} = \sqrt{50 - 50\cos 2} = 5\sqrt{2(1-\cos 2)}.$$

Solutions for Section 6.7

1. (a) Tracing along the graph in Figure 6.28, we see that the approximations for the two solutions are

$$t_1 \approx 1.88 \qquad \text{and} \qquad t_2 \approx 4.41.$$

Note that the first solution, $t_1 \approx 1.88$, is in the second quadrant and the second solution, $t_2 \approx 4.41$, is in the third quadrant. We know that the cosine function is negative in those two quadrants. You can check the two solutions by substituting them into the equation:

$$\cos 1.88 \approx -0.304 \qquad \text{and} \qquad \cos 4.41 \approx -0.298,$$

both of which are close to -0.3.

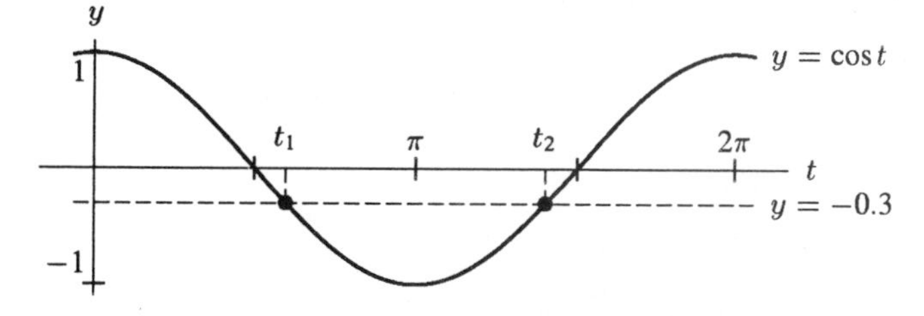

***Figure 6.28**:* The angles t_1 and t_2 are the two solutions to $\cos t = -0.3$ for $0 \le t \le 2\pi$

(b) If your calculator is in radian mode, you should find

$$\cos^{-1}(-0.3) \approx 1.875,$$

which is one of the values we found in part (a) by using a graph. Using the $\boxed{\cos^{-1}}$ key gives only one of the solutions to a trigonometric equation. We find the other solutions by using the symmetry of the unit circle. Figure 6.29 shows that if $t_1 \approx 1.875$ is the first solution, then the second solution is

$$\begin{aligned} t_2 &= 2\pi - t_1 \\ &\approx 2\pi - 1.875 \approx 4.408. \end{aligned}$$

Thus, the two solutions are $t \approx 1.88$ and $t \approx 4.41$.

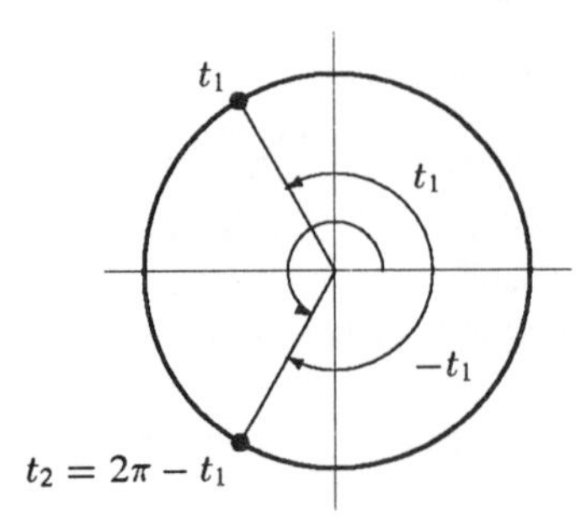

Figure 6.29: By the symmetry of the unit circle, $t_2 = 2\pi - t_1$

5. (a) The reference angle for 120° is $180° - 120° = 60°$, so $\cos 120° = -\cos 60° = -1/2$.
   (b) The reference angle for 135° is $180° - 135° = 45°$, so $\sin 135° = \sin 45° = \sqrt{2}/2$.
   (c) The reference angle for 225° is $225° - 180° = 45°$, so $\cos 225° = -\cos 45° = -\sqrt{2}/2$.
   (d) The reference angle for 300° is $360° - 300° = 60°$, so $\sin 300° = -\sin 60° = -\sqrt{3}/2$.

9. Graph $y = \cos t$ on $0 \leq t \leq 2\pi$ and locate the two points with y-coordinate -0.24. The t-coordinates of these points are approximately $t = 1.8$ and $t = 4.5$. See Figure 6.30.

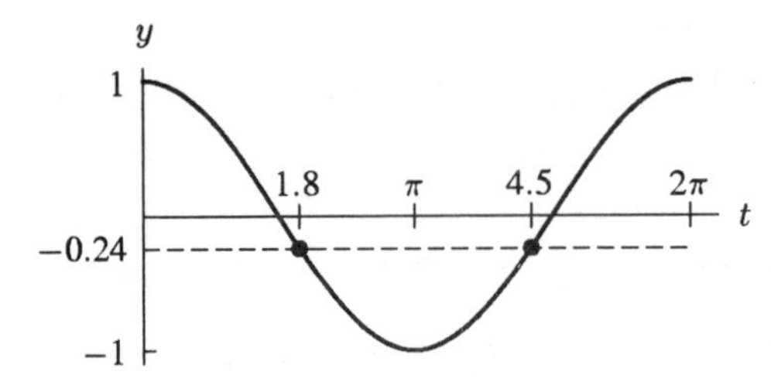

Figure 6.30

13. To solve

$$\tan t = \frac{1}{\tan t}$$

we multiply both sides of the equation by $\tan t$. Multiplication gives us

$$\tan^2 t = 1 \quad \text{or} \quad \tan t = \pm 1.$$

From Figure 6.31, we see that there are two solutions for $\tan t = 1$, and two solutions for $\tan t = -1$, they are approximately $t = 0.79$, $t = 3.93$, and $t = 2.36$, $t = 5.50$.

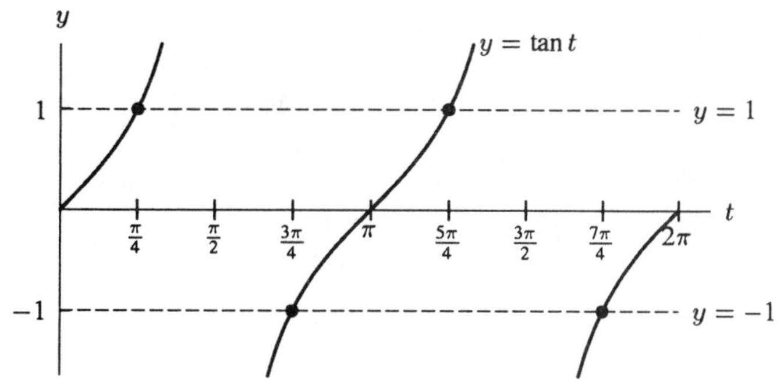

Figure 6.31

To find exact solutions, we have $t = \arctan(\pm 1) = \pm\pi/4$. There are other angles that have a tan of ± 1, namely $\pm 3\pi/4$. So $t = \pi/4, 3\pi/4, 5\pi/4$, and $7\pi/4$ are the solutions in the interval from 0 to 2π.

17. One solution is $\theta = \sin^{-1}(-\sqrt{2}/2) = -\pi/4$, and a second solution is $5\pi/4$, since $\sin(5\pi/4) = -\sqrt{2}/2$. All other solutions are found by adding integer multiples of 2π to these two solutions. See Figure 6.32.

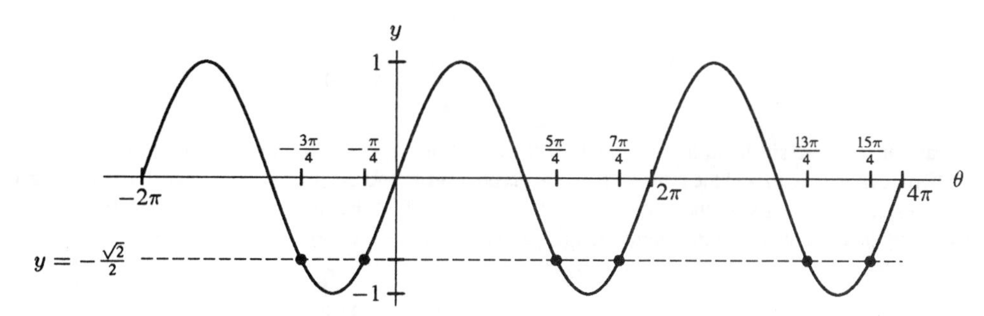

Figure 6.32

21. (a) Let t be the time in hours since 12 noon. Let $d = f(t)$ be the depth in feet in Figure 6.33.

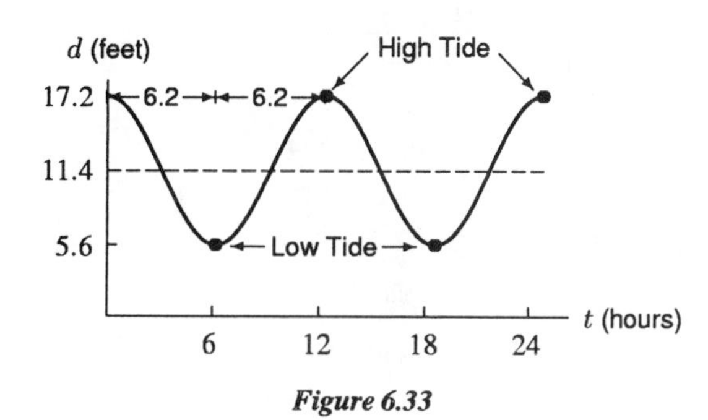

Figure 6.33

(b) The midline is $d = \dfrac{17.2 + 5.6}{2} = 11.4$ and the amplitude is $17.2 - 11.4 = 5.8$. The period is 12.4. Thus we get $d = f(t) = 11.4 + 5.8\cos\left(\dfrac{\pi}{6.2}t\right)$.

(c) We find the first t value when $d = f(t) = 8$:

$$8 = 11.4 + 5.8\cos\left(\frac{\pi}{6.2}t\right)$$

Using the $\cos^{-1}$ function

$$\begin{aligned}\frac{-3.4}{5.8} &= \cos\left(\frac{\pi}{6.2}t\right)\\ \cos^{-1}\left(\frac{-3.4}{5.8}\right) &= \frac{\pi}{6.2}t\\ t &= \frac{6.2}{\pi}\cos^{-1}\left(\frac{-3.4}{5.8}\right) \approx 4.336 \text{ hours.}\end{aligned}$$

Since $0.336(60) \approx 20$ minutes, the latest time the boat can set sail is 4:20 pm.

25.

$$\begin{aligned}\sec^2\alpha + 3\tan\alpha &= \tan\alpha\\ 1 + \tan^2\alpha + 3\tan\alpha &= \tan\alpha\\ \tan^2\alpha + 2\tan\alpha + 1 &= 0\\ (\tan\alpha + 1)^2 &= 0\\ \tan\alpha &= -1\\ \alpha &= \frac{3\pi}{4}, \frac{7\pi}{4}\end{aligned}$$

29. (a) Graph $y = 3 - 5\sin 4t$ on the interval $0 \le t \le \pi/2$, and locate values where the function crosses the t-axis. Alternatively, we can find the points where the graph $5\sin 4t$ and the line $y = 3$ intersect. By looking at the graphs of these two functions on the interval $0 \le t \le \pi/2$, we find that they intersect twice. By zooming in we can identify these points of intersection as roughly $t_1 \approx 0.16$ and $t_2 \approx 0.625$. See Figure 6.34.

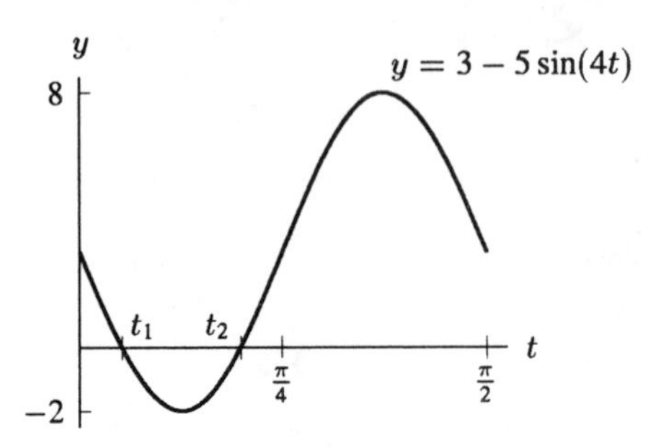

Figure 6.34

(b) Solve for $\sin(4t)$ and then use arcsin:

$$\begin{aligned}5\sin(4t) &= 3\\ \sin(4t) &= \frac{3}{5}\\ 4t &= \arcsin\left(\frac{3}{5}\right).\end{aligned}$$

So $t_1 = \dfrac{\arcsin(3/5)}{4} \approx 0.161$ is a solution. But the angle $\pi - \arcsin(3/5)$ has the same sine as $\arcsin(3/5)$. Solving $4t = \pi - \arcsin(3/5)$ gives $t_2 = \dfrac{\pi}{4} - \dfrac{\arcsin(3/5)}{4} \approx 0.625$ as a second solution.

33. (a) $\cos^{-1}\left(\frac{1}{2}\right)$ is the angle between 0 and π whose cosine is $1/2$. Since $\cos\left(\frac{\pi}{3}\right) = 1/2$, we have $\cos^{-1}\left(\frac{1}{2}\right) = \pi/3$.
   (b) Similarly, $\cos^{-1}\left(-\frac{1}{2}\right)$ is the angle between 0 and π whose cosine is $-1/2$. Since $\cos\left(\frac{2\pi}{3}\right) = -1/2$, we have $\cos^{-1}\left(-\frac{1}{2}\right) = 2\pi/3$.
   (c) From the first part, we saw that $\cos^{-1}\left(\frac{1}{2}\right) = \pi/3$. This gives

$$\cos\left(\cos^{-1}\left(\frac{1}{2}\right)\right) = \cos\left(\frac{\pi}{3}\right) = \frac{1}{2}.$$

This is not at all surprising; after all, what we are saying is that the cosine of the inverse cosine of a number is that number. However, the situation is not as straightforward as it may appear. The next part of this question illustrates the problem.
   (d) The cosine of $5\pi/3$ is $1/2$. And since $\cos^{-1}\left(\frac{1}{2}\right) = \pi/3$, we have

$$\cos^{-1}\left(\cos\left(\frac{5\pi}{3}\right)\right) = \cos^{-1}\left(\frac{1}{2}\right) = \frac{\pi}{3}.$$

Thus, we see that the inverse cosine of the cosine of an angle does not necessarily equal that angle.

37. (a) In Figure 6.35, the earth's center is labeled O and two radii are extended, one through S, your ship's position, and one through H, the point on the horizon. Your line of sight to the horizon is tangent to the surface of the earth. A line tangent to a circle at a given point is perpendicular to the circle's radius at that point. Thus, since your line of sight is tangent to the earth's surface at H, it is also perpendicular to the earth's radius at H. This means that triangle OCH is a right triangle. Its hypotenuse is $r + x$ and its legs are r and d. From the Pythagorean theorem, we have

$$\begin{aligned} r^2 + d^2 &= (r+x)^2 \\ d^2 &= (r+x)^2 - r^2 \\ &= r^2 + 2rx + x^2 - r^2 = 2rx + x^2. \end{aligned}$$

Since d is positive, we have $d = \sqrt{2rx + x^2}$.

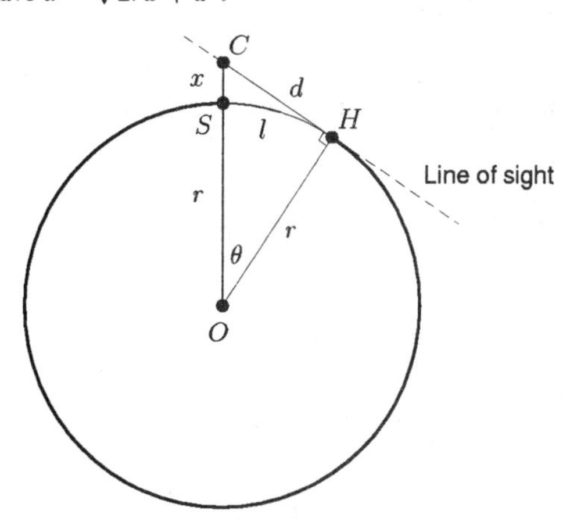

Figure 6.35

   (b) We begin by using the formula obtained in part (a).

$$\begin{aligned} d &= \sqrt{2rx + x^2} \\ &= \sqrt{2(6{,}370{,}000)(50) + 50^2} \\ &\approx 25{,}238.9. \end{aligned}$$

Thus, you would be able to see a little over 25 kilometers from the crow's nest C.

Having found a formula for d, we will now try to find a formula for l, the distance along the earth's surface from the ship to the horizon H. In Figure 6.35, l is the arc length specified by the angle θ (in radians). The formula for arc length is

$$l = r\theta.$$

In this case, we must determine θ. From Figure 6.35 we see that

$$\cos\theta = \frac{\text{adjacent}}{\text{hypotenuse}} = \frac{r}{r+x}.$$

Thus,

$$\theta = \cos^{-1}\left(\frac{r}{r+x}\right)$$

since $0 \leq \theta \leq \pi/2$. This means that

$$\begin{aligned} l = r\theta &= r\cos^{-1}\left(\frac{r}{r+x}\right) \\ &= 6{,}370{,}000\cos^{-1}\left(\frac{6{,}370{,}000}{6{,}370{,}050}\right) \approx 25{,}238.8 \text{ meters.} \end{aligned}$$

There is very little difference—about 0.1 m or 10 cm—between the distance d that you can see and the distance l that the ship must travel to reach the horizon. If this is surprising, keep in mind that Figure 6.35 has not been drawn to scale. In reality, the mast height x is significantly smaller than the earth's radius r so that the point C in the crow's nest is very close to the ship's position at point S. Thus, the line segment d and the arc l are almost indistinguishable.

Solutions for Chapter 6 Review

1. True, because $\sin x$ is an odd function.
5. True, by the sum of angles identity.
9. False, by the sum of angles identity.
13. (a) Since $1.57 < 2 < 3.14$, you will be in the quadrant II.
    (b) Since $3.14 < 4 < 4.71$, you will be in the quadrant III.
    (c) Since $4.71 < 6 < 6.28$, you will be in the quadrant IV.
    (d) Since $0 < 1.5 < 1.57$, you will be in the quadrant I.
    (e) Since $3.14 < 3.2 < 4.71$, you will be in the quadrant III.
17. (a) $\pi/3 = 1.047197551$
    (b) $1/\sqrt{2} = 0.7071067812$
    (c) $\pi/6 = 0.5235987756$
    (d) $\pi/4 = 0.7853981634$
    (e) $\sqrt{3}/2 = 0.8660254038$
    (f) $\pi/2 = 1.570796327$
21. Graph $y = \cos(4\theta)$ and $y = -1$ and find the θ-coordinates of the intersection points. There are four solutions: $\theta \approx \pi/4$, $\theta \approx 3\pi/4$, $\theta \approx 5\pi/4$, and $\theta \approx 7\pi/4$. See Figure 6.36.

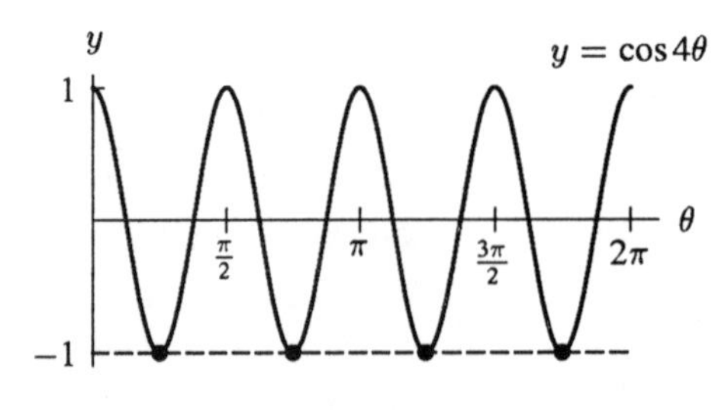

Figure 6.36

25.
$$\begin{aligned}\sin(2\alpha)+3&=4\\ \sin(2\alpha)&=1\\ 2\alpha&=\frac{\pi}{2},\ \frac{5\pi}{2}\\ \alpha&=\frac{\pi}{4},\ \frac{5\pi}{4}.\end{aligned}$$

29.
$$\begin{aligned}3\cos^2\alpha+2&=3-2\cos\alpha\\ 3\cos^2\alpha+2\cos\alpha-1&=0\\ (3\cos\alpha-1)(\cos\alpha+1)&=0\end{aligned}$$

$$\begin{aligned}3\cos\alpha-1&=0 & \cos\alpha+1&=0\\ \cos\alpha&=\tfrac{1}{3} & \cos\alpha&=-1\\ \alpha&=1.2310,\ 5.0522 & \alpha&=\pi\end{aligned}$$

33. (a) The two beams form the legs of a right triangle with a 45° angle inside the unit circle. The bottom leg, which is the distance from the center to the beam intersection, is $\cos(45°)=\sqrt{2}/2$. Thus the distance of the sensor S from the point of intersection is
$$1-\frac{\sqrt{2}}{2}=\frac{2-\sqrt{2}}{2}\text{ meters.}$$

(b)

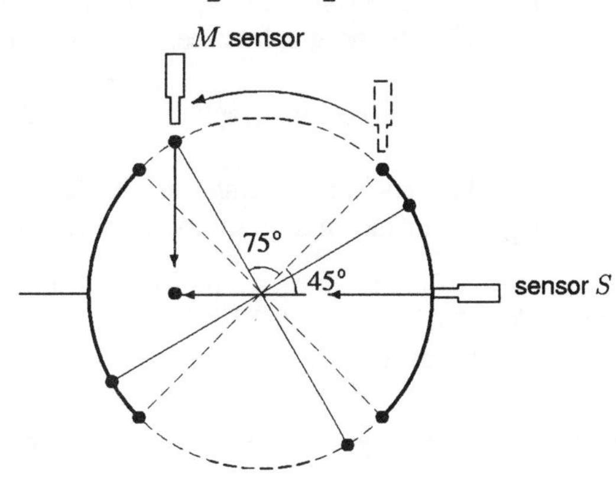

Figure 6.37

The new position of the revolving door is shown in Figure 6.37. The panel with the sensor M now makes an angle of $45°+75°=120°$ with the positive x-axis. The x-coordinate is $\cos 120°=-1/2$. Thus the distance of the wall sensor from the point of intersection is $1-(-1/2)=3/2$ meters.

37. The circumference of the outer edge is
$$6(2\pi)=12\pi\text{ cm.}$$
A point on the outer edge travels 100 times this distance in one minute. Thus, a point on the outer edge must travel at the speed of 1200π cm/minute or roughly 3770 cm/min.

The circumference of the inner edge is
$$0.75(2\pi)=1.5\pi\text{ cm.}$$
A point on the inner edge travels 100 times this distance in one minute. Thus, a point on the inner edge must travel at the speed of 150π cm/minute or roughly 471 cm/min.

41. (a) We are looking for the graph of a function with amplitude one but a period of π; only $C(t)$ qualifies.
    (b) We are looking for the graph of a function with amplitude one and period 2π but which is shifted up by two units; only $D(t)$ qualifies.
    (c) We are looking for the graph of a function with amplitude 2 and period 2π; only $A(t)$ qualifies.
    (d) Only $B(t)$ is left and we are looking for the graph of a function with amplitude one and period 2π but which has been shifted to the left by two units. This checks with $B(t)$.

45. $f(t) = 150 + 150\sin\left(\frac{\pi}{10}t - \frac{\pi}{2}\right)$

49.

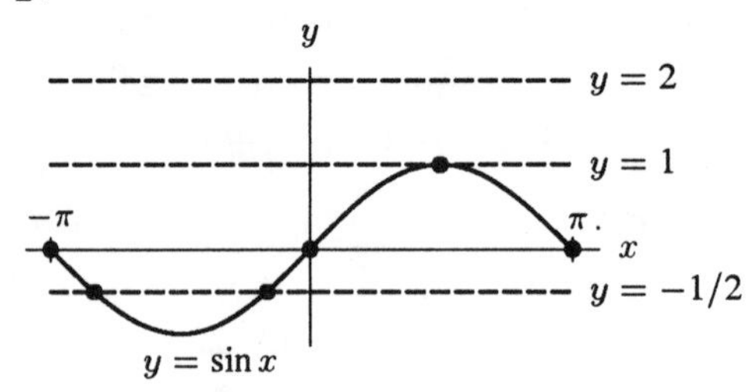

Figure 6.38

Figure 6.38 shows the graph of $y = \sin x$ for $-\pi \leq x \leq \pi$, together with the horizontal lines $y = 1$, $y = -1/2$, $y = 2$ and $y = 0$ (the x-axis).

(a) Since the line $y = 1$ cuts the graph once, the equation $\sin x = 1$ has one solution for $-\pi \leq x \leq \pi$.
(b) Since the line $y = -1/2$ cuts the graph twice, the equation $\sin x = -1/2$ has two solutions for $-\pi \leq x \leq \pi$.
(c) Since the line $y = 0$ (the x-axis) cuts the graph three times, the equation $\sin x = 0$ has three solutions for $-\pi \leq x \leq \pi$.
(d) Since the line $y = 2$ does not cut the graph at all, the equation $\sin x = 2$ has no solutions for $-\pi \leq x \leq \pi$. (In fact, this equation has no solutions for any other x-values either.)

53. (a) The period of the first timer is 3 minutes; the period of the second is 4 minutes.
    (b) The first timer beeps at 9:03, 9:06, 9:09, 9:12, 9:15, 9:18, ... The second timer beeps at 9:03, 9:07, 9:11, 9:15, 9:19, ... So the next time they beep together is 9:15 am.
    (c) The period of beeping together is 12 minutes.

57. Since the temperature, y, repeats each year, we use a trigonometric function of the form

$$y = A\sin(B(x - h)) + k.$$

Since the period is 12 months, $2\pi/B = 12$, so $B = \pi/6$. The midline k is the equilibrium or average value, so

$$k \approx (31 + 19)/2 = 25.$$

The amplitude is approximated by

$$A \approx (31 - 19)/2 = 6.$$

Since the temperature crosses the midline in April ($t = 3$, if $t = 0$ is Jan 1), the graph is shifted 3 to the right, so $h = 3$. Thus,

$$y = 6\sin\left(\frac{\pi}{6}(t - 3)\right) + 25.$$

CHAPTER SEVEN

Solutions for Section 7.1

1. By the Pythagorean theorem, the hypotenuse has length $\sqrt{1^2+2^2}=\sqrt{5}$.
   (a) $\tan\theta = \dfrac{\text{opposite}}{\text{adjacent}} = \dfrac{2}{1} = 2.$
   (b) $\sin\theta = \dfrac{\text{opposite}}{\text{hypotenuse}} = \dfrac{2}{\sqrt{5}}.$
   (c) $\cos\theta = \dfrac{\text{adjacent}}{\text{hypotenuse}} = \dfrac{1}{\sqrt{5}}.$

5. From the figure in the text, we see that
$$\sin x = \frac{0.83}{1},$$
so
$$x = \sin^{-1}(0.83).$$
Using a calculator, we find that $x = \sin^{-1}(0.83) \approx 0.979$.

9. See Figure 7.1. The angle θ is the sun's angle of elevation. Here, $\tan\theta = \dfrac{50}{60} = \dfrac{5}{6}$. So, $\theta = \tan^{-1}\left(\dfrac{5}{6}\right) \approx 39.8°$.

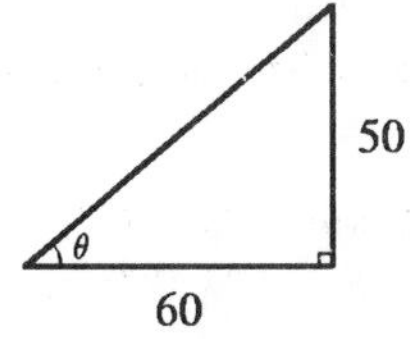

Figure 7.1

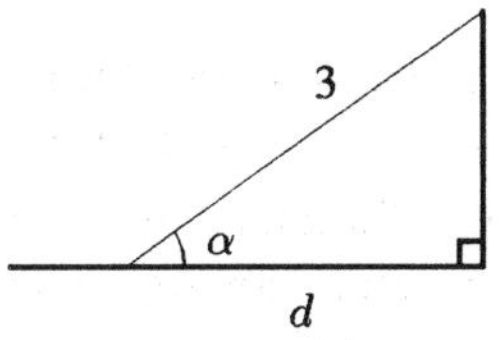

Figure 7.2

13. Let d be the distance from the base of the ladder to the wall; see Figure 7.2. Then, $d/3 = \cos\alpha$, so $d = 3\cos\alpha$ meters.

17. (a) Since $\sin 45° = \dfrac{h}{125}$, we have $h = 125\sin 45° \approx 88.39$ feet.
    (b) Since $\sin 30° = \dfrac{h}{125}$, we have $h = 125\sin 30° = 62.5$ feet.
    (c) Since $\cos 45° = \dfrac{c}{125}$, we have $c = 125\cos(45°) \approx 88.39$ feet.
    Since $\cos 30° = \dfrac{d}{125}$, we have $d \approx 108.25$ feet.

Solutions for Section 7.2

1. By the Law of Sines, we have
$$\frac{x}{\sin 100°} = \frac{6}{\sin 18°}$$
$$x = 6\left(\frac{\sin 100°}{\sin 18°}\right) \approx 19.12.$$

5. In Figure 7.3, we have

$$\theta = 180° - 90° - 10° \qquad a = 12\cos 10° \qquad b = 12\sin 10°$$
$$\theta = 80° \qquad a \approx 12(0.985) \qquad b \approx 12(0.174)$$
$$a \approx 11.82. \qquad b \approx 2.08.$$

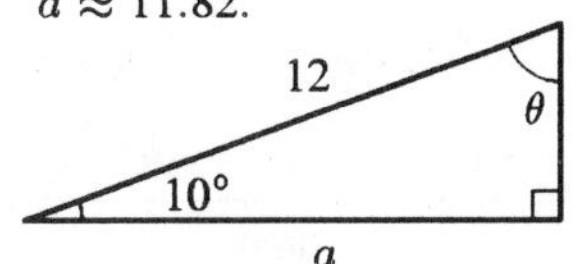

Figure 7.3

9. In Figure 7.4, use the Law of Sines:

$$\frac{\sin(30°)}{259} = \frac{\sin\beta}{510}$$

to obtain $\sin\beta \approx 0.9846$ and use $\sin^{-1}$ to find $\beta_1 \approx 79.9$ or $\beta_2 \approx 100.1$. We then know $\alpha_1 = 180° - 30° - 79.9° \approx 70.1°$, or $\alpha_2 = 180° - 30° - 100.1° \approx 49.9°$. We can use the value of α and the Law of Sines to find the length of side a:

$$\frac{a_1}{\sin(70.1°)} = \frac{259}{\sin 30°}, \quad \text{or} \quad \frac{a_2}{\sin(49.9°)} = \frac{259}{\sin 30°}.$$
$$a_1 \approx 487.07 \text{ ft} \qquad a_2 \approx 396.23 \text{ ft}$$

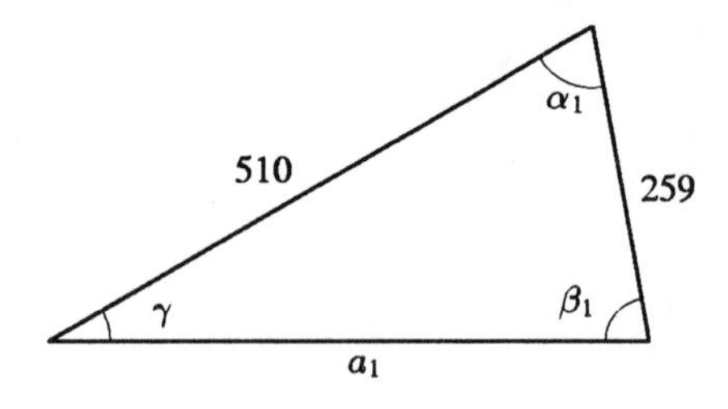

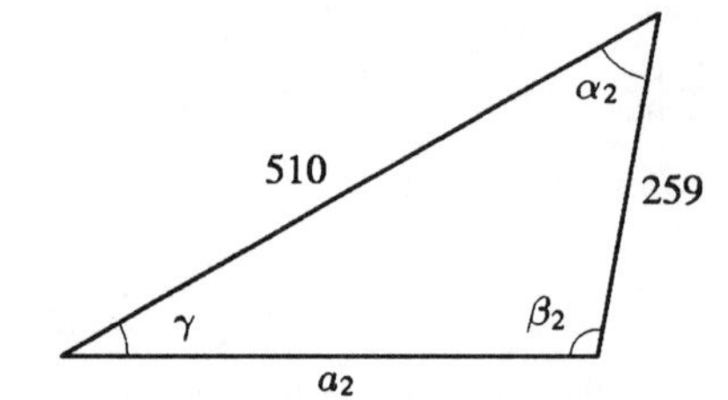

Figure 7.4

13.

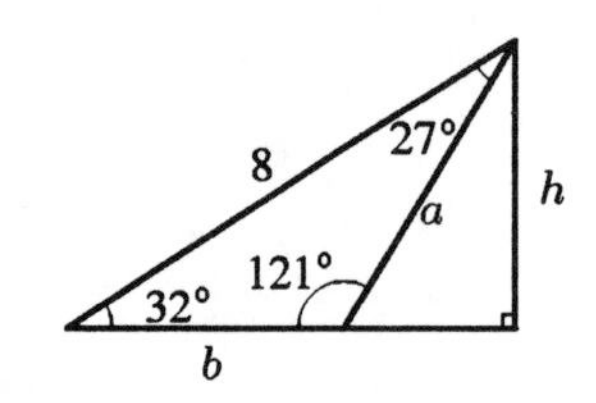

Figure 7.5

(a) $\frac{\sin 121°}{8} = \frac{\sin 32°}{a}$, so $a = \frac{8\sin 32°}{\sin 121°} \approx 4.95$. Similarly, $b = \frac{8\sin 27°}{\sin 121°} \approx 4.24$.

(b) Construct an altitude h as in Figure 7.5. We have $\sin 32° = \frac{h}{8}$, so $h = 8\sin 32° \approx 4.24$. Then area of the triangle is $\frac{1}{2}(8)(4.24) \approx 17$.

17. From the figure, we see that $\sin A = \frac{h}{b}$, which gives $h = b\sin A$. We also have $\sin B = \frac{h}{a}$, which gives $h = a\sin B$. Thus, $b\sin A = a\sin B$, which gives the Law of Sines:

$$\frac{\sin A}{a} = \frac{\sin B}{b}.$$

21. In Figure 7.6, the fire stations are at A and B and the forest fire is at C. The angle at C is $180° - 54° - 58° = 68°$.

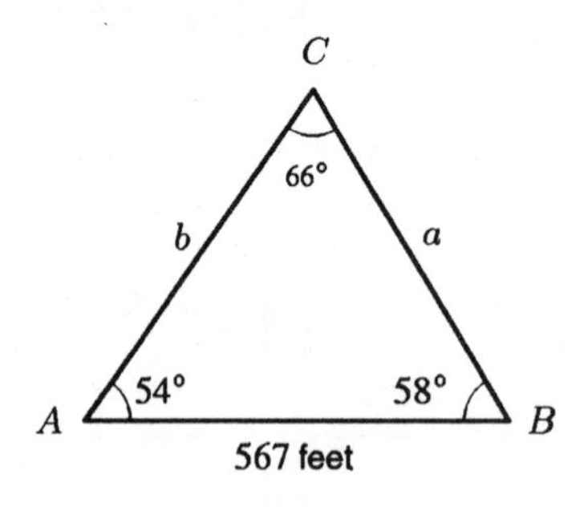

Figure 7.6

Solving for a and b using the Law of Sines, we get

$$\frac{567}{\sin 68°} = \frac{a}{\sin 54°} \qquad \frac{567}{\sin 68°} = \frac{b}{\sin 58°}$$

$$a = \frac{567 \sin 54°}{\sin 68°} \qquad b = \frac{567 \sin 58°}{\sin 68°}$$

$$a = 494.74 \qquad b = 518.61$$

The fire station at point B is closer by $518.61 - 494.74 = 23.87$ feet.

25. From Figure 7.7, we see $\angle ABC = 180° - 93° - 49° = 38°$. Using the Law of Sines, we have

$$\frac{102}{\sin 38°} = \frac{c}{\sin 49°}$$
$$c = \frac{102 \sin 49°}{\sin 38°} = 125.04 \text{ feet.}$$

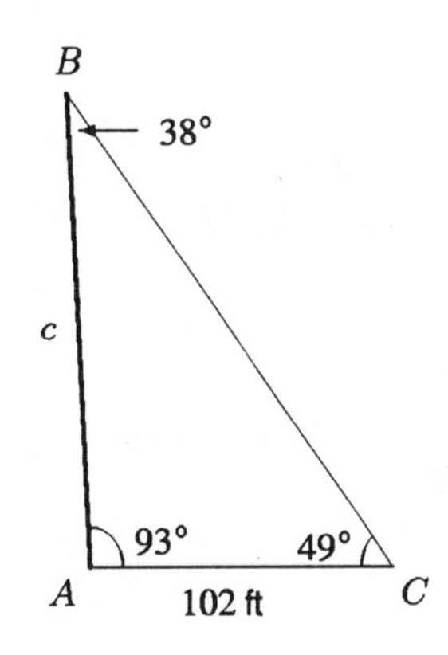

Figure 7.7

29. One way to organize this situation is to use the abbreviations from high school geometry. The six possibilities are { SSS, SAS, SSA, ASA, AAS, AAA }.

SSS Knowing all three sides allows us to find the angles by using the Law of Cosines.

SAS Knowing two sides and the included angle allows us to find the third side length by using the Law of Cosines. We can then use the SSS procedure.

SSA Knowing two sides but not the included angle is called the ambiguous case, because there could be two different solutions. Use the Law of Sines to find one of the missing angles, which, because we use the arcsin, may give two values. Or, use the Law of Cosines, which produces a quadratic equation that may also give two values. Treating these cases separately we can continue to find all sides and angles using the SAS procedure.

ASA Knowing two angles allows us to easily find the third angle. Use the Law of Sines to find each side.

AAS Find the third angle and then use the Law of Sines to find each side.

AAA This has an infinite number of solutions because of similarity of triangles. Once one side is known, then the ASA or AAS procedure can be followed.

Solutions for Section 7.3

1. Since $\cos 2t$ has period π and $\sin t$ has period 2π, if the result we want holds for $0 \le t \le 2\pi$, it holds for all t. So let's concentrate on the interval $0 \le t \le 2\pi$.

   Solving $\cos 2t = 0$ gives $t = \pi/4, 3\pi/4, 5\pi/4, 7\pi/4$.

   From the graph in Figure 7.8, we see $\cos 2t > 0$ for $0 \le t < \pi/4$, $3\pi/4 < t < 5\pi/4$, $7\pi/4 < t \le 2\pi$.

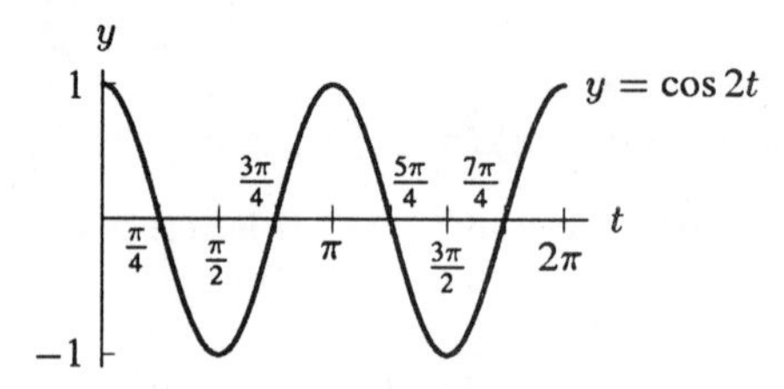

Figure 7.8

Solving $1 - 2\sin^2 t = 0$ gives

$$\sin^2 t = \frac{1}{2}$$
$$\sin t = \pm\frac{1}{\sqrt{2}},$$

so $t = \pi/4, 3\pi/4, 5\pi/4, 7\pi/4$.

From the graph of $y = \sin t$ and the lines $y = 1/\sqrt{2}$ and $y = -1/\sqrt{2}$ in Figure 7.9, we see that $-1/\sqrt{2} < \sin t < 1/\sqrt{2}$ on the same intervals that $\cos 2t > 0$.

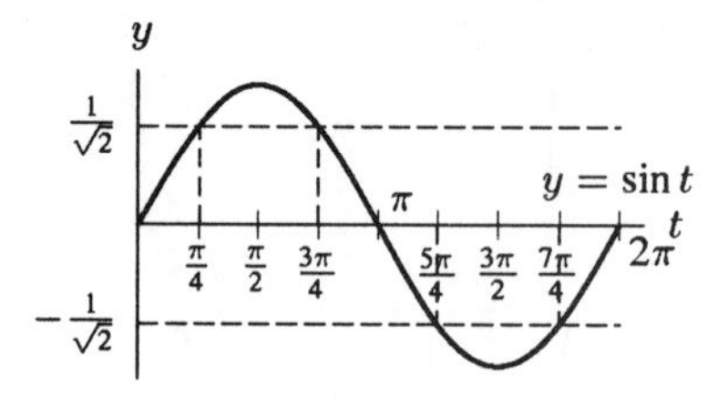

Figure 7.9

Now if $-1/\sqrt{2} < \sin t < 1/\sqrt{2}$, then

$$\sin^2 t < \frac{1}{2}$$
$$1 - 2\sin^2 t > 0.$$

Thus, $\cos 2t$ and $1 - 2\sin^2 t$ have the same sign for all t.

5. Both functions are symmetric about the y-axis. (They are even functions.) They are both equal to one when $x = 0$. They both have an amplitude of one. However, $\cos(2x)$ is periodic, while $\cos(x^2)$ is not. See Figure 7.10.

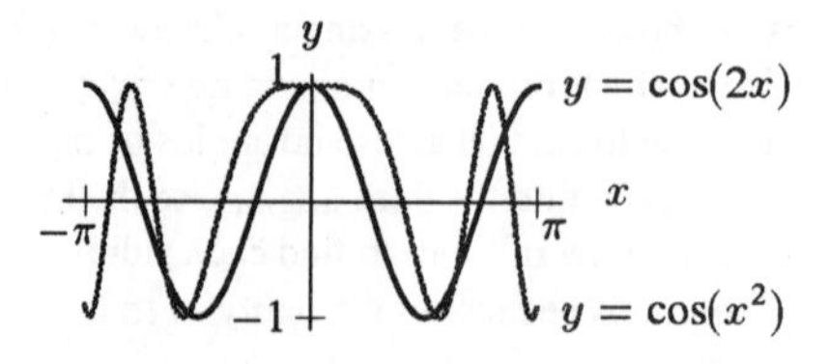

Figure 7.10

9. By graphing we can determine identities. The graphs all show the same window, $-2\pi \le x \le 2\pi$, $-4 \le y \le 4$.

(a)
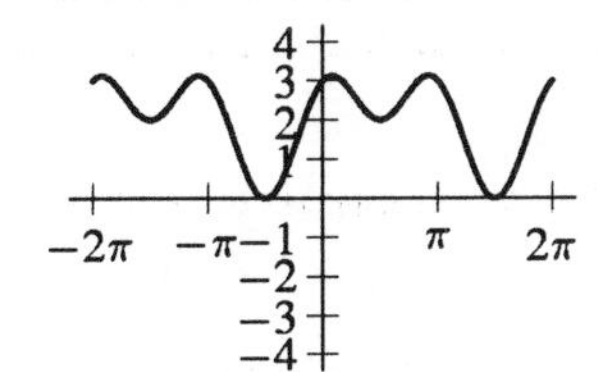

(b)
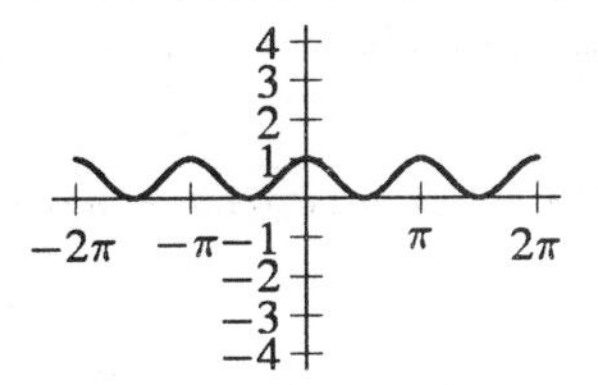

(c)
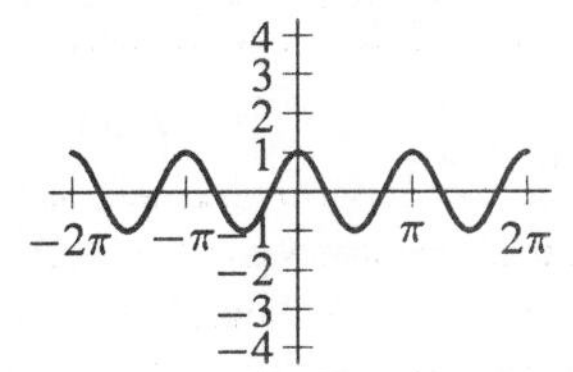

(d)
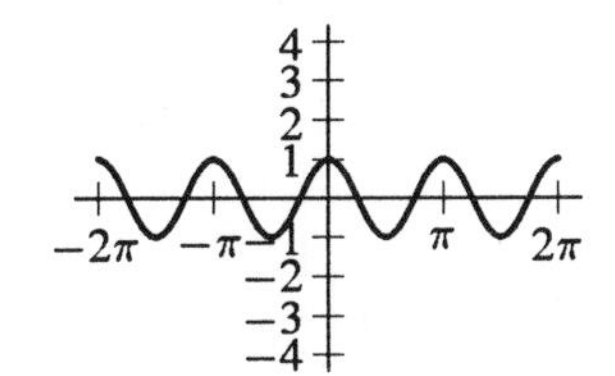

(e)
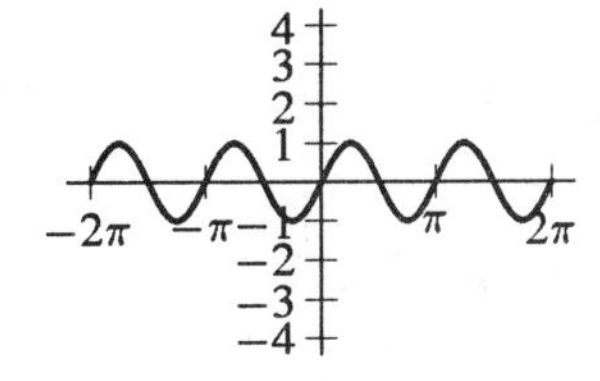

(f)
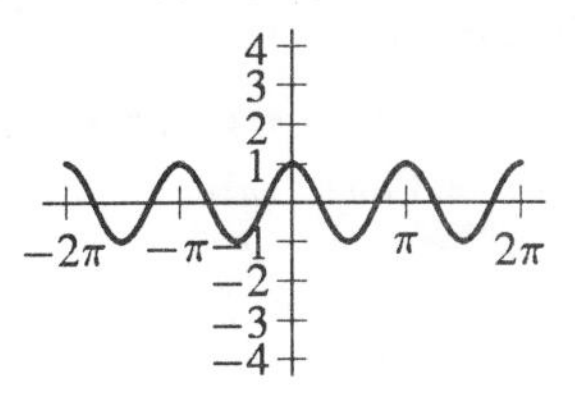

(g)
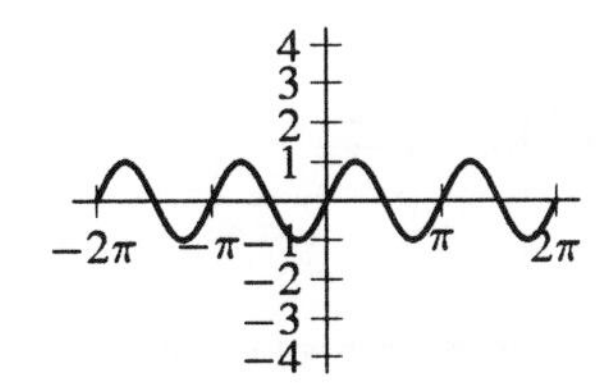

(h)
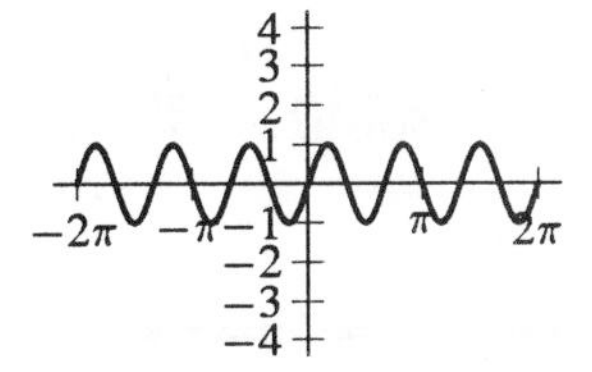

(i)
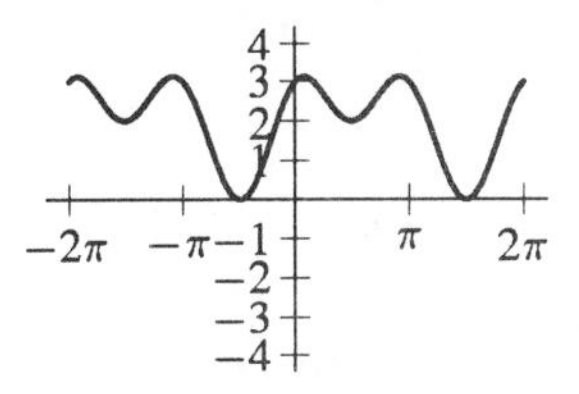

(j)
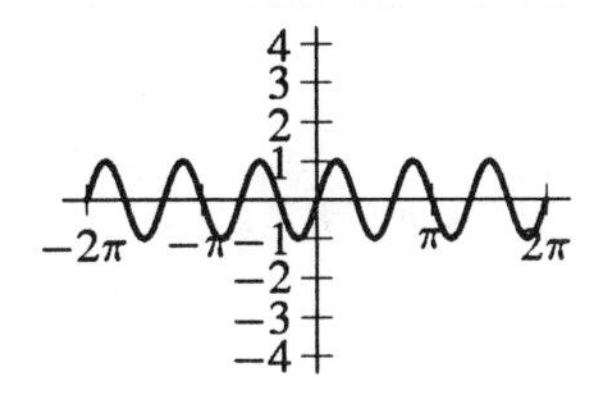

(k)
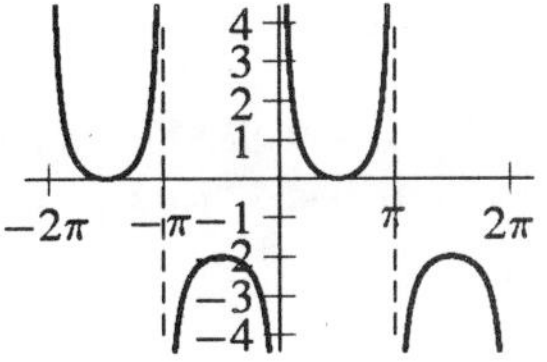

(l)
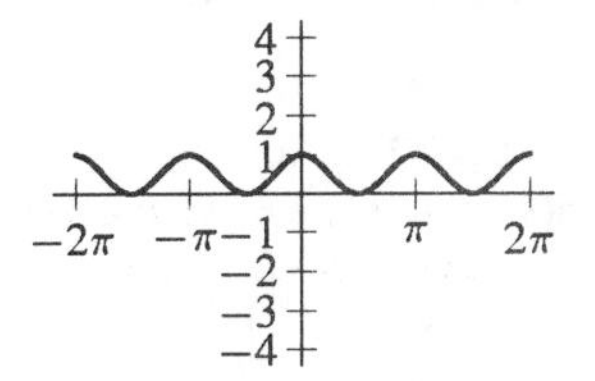

(m)
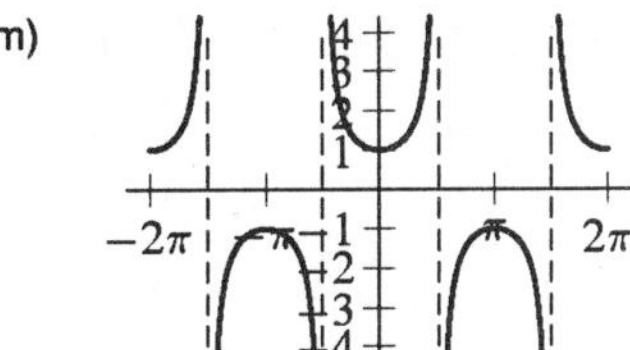

The following pairs of expressions are identical: a and i; b and l; c and d and f; e and g; h and j. Note that k and m are different functions. We can verify the identities algebraically. For example, a and i: $2\cos^2 t + \sin t + 1 = 2(1 - \sin^2 t) + \sin t + 1 = -2\sin^2 t + \sin t + 3$.

13. Not an identity. False for $x = 2$.

17. Identity. $\dfrac{\sin 2x}{1 + \cos 2x} = \dfrac{2\sin x \cos x}{1 + 2\cos^2 x - 1} = \dfrac{2\sin x \cos x}{2\cos^2 x} = \dfrac{\sin x}{\cos x} = \tan x.$

21. Get a common denominator:

$$\begin{aligned}
\frac{\cos x}{1 - \sin x} - \tan x &= \frac{\cos x}{1 - \sin x} - \frac{\sin x}{\cos x} \\
&= \frac{\cos^2 x - \sin x(1 - \sin x)}{(1 - \sin x)(\cos x)} \\
&= \frac{\cos^2 x - \sin x + \sin^2 x}{(1 - \sin x)(\cos x)} \\
&= \frac{1 - \sin x}{(1 - \sin x)\cos x} = \frac{1}{\cos x}.
\end{aligned}$$

25. (a) $\cos\theta = x$.
(b) $\cos\left(\frac{\pi}{2} - \theta\right) = \sin\theta = \sqrt{1-\cos^2\theta} = \sqrt{1-x^2}$.
(c) $\tan^2\theta = \left(\frac{\sin\theta}{\cos\theta}\right)^2 = \left(\frac{\sqrt{1-x^2}}{x}\right)^2 = \frac{1-x^2}{x^2}$.
(d) $\sin(2\theta) = 2\sin\theta\cos\theta = 2\sqrt{1-x^2}(x)$.
(e) $\cos(4\theta) = 1 - 2\sin^2(2\theta)$. Now we use part (d):
$\cos(4\theta) = 1 - 2(2x\sqrt{1-x^2})^2 = 1 - 8x^2(1-x^2)$.
(f) $\sin(\cos^{-1}x) = \sin(\theta) = \sqrt{1-x^2}$.

29. We have $\sin\theta = \frac{x+1}{5}$, so $\cos 2\theta = 1 - 2\sin^2\theta = 1 - 2\left(\frac{x+1}{5}\right)^2 = 1 - \frac{2(x+1)^2}{25}$.

33. (a) Let $x = 3\sin u$. Then

$$f(x) = \sqrt{9-(3\sin u)^2} = \sqrt{9\cos^2 u} = 3\cos u = 3\left|\cos\left(\sin^{-1}\left(\frac{x}{3}\right)\right)\right|.$$

(b) Let $x = \sqrt{5}\sin u$. Then

$$g(x) = \sqrt{5-(\sqrt{5}\sin u)^2} = \sqrt{5cos^2u} = \sqrt{5}\cos u = \sqrt{5}\left|\cos\left(\sin^{-1}\left(\frac{x}{\sqrt{5}}\right)\right)\right|.$$

Solutions for Section 7.4

1. (a) $\sin(15+42) = \sin 15\cos 42 + \sin 42\cos 15 = 0.839$. See Figure 7.11.
(b) $\sin(15-42) = \sin 15\cos 42 - \sin 42\cos 15 = -0.454$. See Figure 7.12.

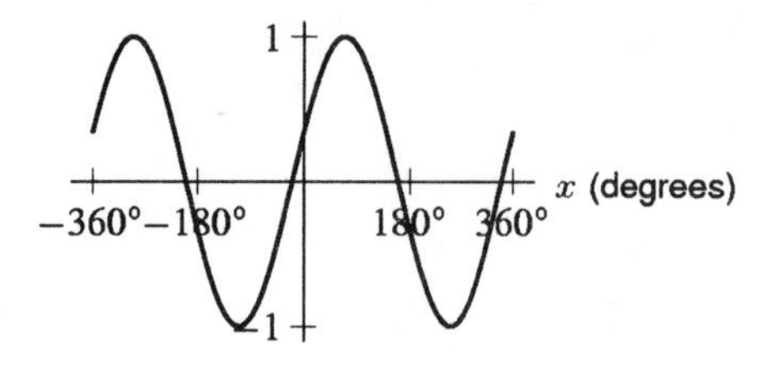

Figure 7.11

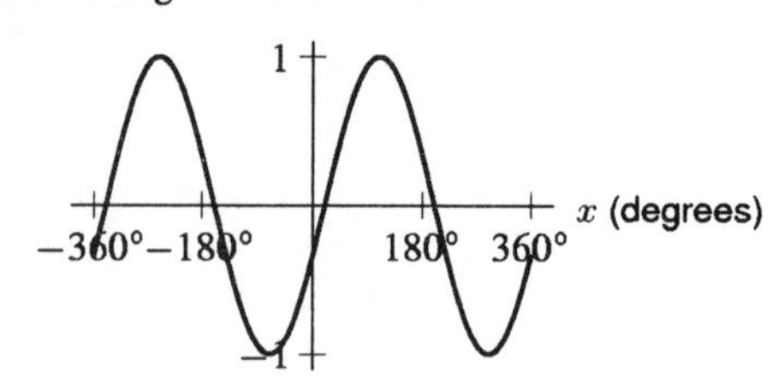

Figure 7.12

(c) $\cos(15+42) = \cos 15\cos 42 - \sin 15\sin 42 = 0.545$. See Figure 7.13.
(d) $\cos(15-42) = \cos 15\cos 42 + \sin 15\sin 42 = 0.891$. See Figure 7.14.

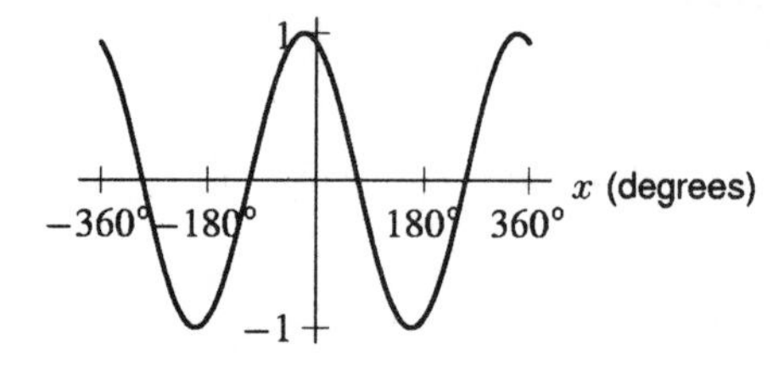

Figure 7.13

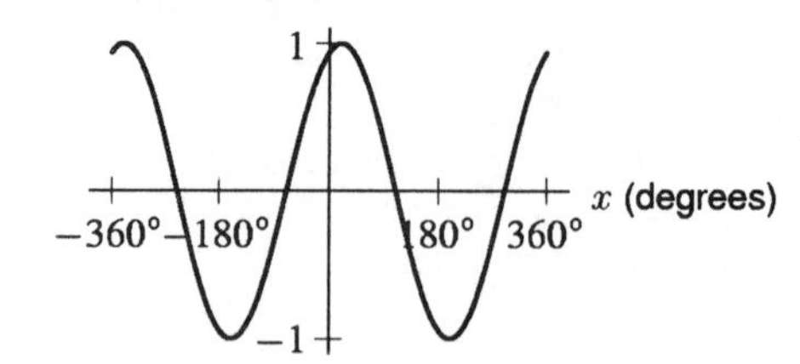

Figure 7.14

5. We have $A = \sqrt{8^2+(-6)^2} = \sqrt{100} = 10$. Since $\cos\phi = 8/10 = 0.8$ and $\sin\phi = -6/10 = -0.6$, we know that ϕ is in the fourth quadrant. Thus,

$$\tan\phi = -\frac{6}{8} = -0.75 \quad \text{and} \quad \phi = \tan^{-1}(-0.75) = -0.64,$$

so $8\sin t - 6\cos t = 10\sin(t - 0.64)$.

9. For the sine, we have $\sin 2t = \sin(t+t) = \sin t\cos t + \sin t\cos t = 2\sin t\cos t$. This is the double angle formula for sine.

   For cosine, we have $\cos 2t = \cos(t+t) = \cos t\cos t - \sin t\sin t = \cos^2 t - \sin^2 t$. This is the double angle formula for cosine.

13. We start with
$$\sin u + \sin v = 2\sin\left(\frac{u+v}{2}\right)\cos\left(\frac{u-v}{2}\right).$$
Since $-\sin v = \sin(-v)$, we can write
$$\begin{aligned}\sin u - \sin v &= \sin u + \sin(-v)\\ &= 2\sin\left(\frac{u+(-v)}{2}\right)\cos\left(\frac{u-(-v)}{2}\right)\\ &= 2\sin\left(\frac{u-v}{2}\right)\cos\left(\frac{u+v}{2}\right)\\ &= 2\cos\left(\frac{u+v}{2}\right)\sin\left(\frac{u-v}{2}\right).\end{aligned}$$

17. Since $0 < \ln x < \frac{\pi}{2}$ and $0 < \ln y < \frac{\pi}{2}$, the angles represented by $\ln x$ and $\ln y$ are in the first quadrant. This means that both their sine and cosine values will be positive. Since $\ln(xy) = \ln x + \ln y$, we can write
$$\sin(\ln(xy)) = \sin(\ln x + \ln y).$$
By the sum-of-angle formula we have
$$\sin(\ln x + \ln y) = \sin(\ln x)\cos(\ln y) + \cos(\ln x)\sin(\ln y).$$
Since cosine is positive, we have
$$\cos(\ln x) = \sqrt{1-\sin^2(\ln x)} = \sqrt{1-\left(\frac{1}{3}\right)^2} = \frac{\sqrt{8}}{3}$$
and
$$\cos(\ln y) = \sqrt{1-\sin^2(\ln y)} = \sqrt{1-\left(\frac{1}{5}\right)^2} = \frac{\sqrt{24}}{5}.$$
Thus,
$$\begin{aligned}\sin(\ln x + \ln y) &= \sin(\ln x)\cos(\ln y) + \cos(\ln x)\sin(\ln y)\\ &= \left(\frac{1}{3}\right)\left(\frac{\sqrt{24}}{5}\right) + \left(\frac{\sqrt{8}}{3}\right)\left(\frac{1}{5}\right)\\ &= \frac{\sqrt{24}+\sqrt{8}}{15}\\ &\approx 0.515160.\end{aligned}$$

21. (a) From $\cos 2u = 2\cos^2 u - 1$, we obtain $\cos u = \pm\sqrt{\dfrac{1+\cos 2u}{2}}$ and letting $u = \frac{v}{2}$, $\cos\dfrac{v}{2} = \pm\sqrt{\dfrac{1+\cos v}{2}}$.

    (b) From $\tan\dfrac{1}{2}v = \dfrac{\sin\frac{1}{2}v}{\cos\frac{1}{2}v} = \dfrac{\pm\sqrt{\frac{1-\cos v}{2}}}{\pm\sqrt{\frac{1+\cos v}{2}}}$ we simplify to get $\tan\dfrac{1}{2}v = \pm\sqrt{\dfrac{1-\cos v}{1+\cos v}}$.

    (c) The sign of $\sin\frac{1}{2}v$ is $+$, the sign of $\cos\frac{1}{2}v$ is $-$, and the sign of $\tan\frac{1}{2}v$ is $-$.

    (d) The sign of $\sin\frac{1}{2}v$ is $-$, the sign of $\cos\frac{1}{2}v$ is $-$, and the sign of $\tan\frac{1}{2}v$ is $+$.

    (e) The sign of $\sin\frac{1}{2}v$ is $-$, the sign of $\cos\frac{1}{2}v$ is $+$, and the sign of $\tan\frac{1}{2}v$ is $-$.

Solutions for Section 7.5

1.

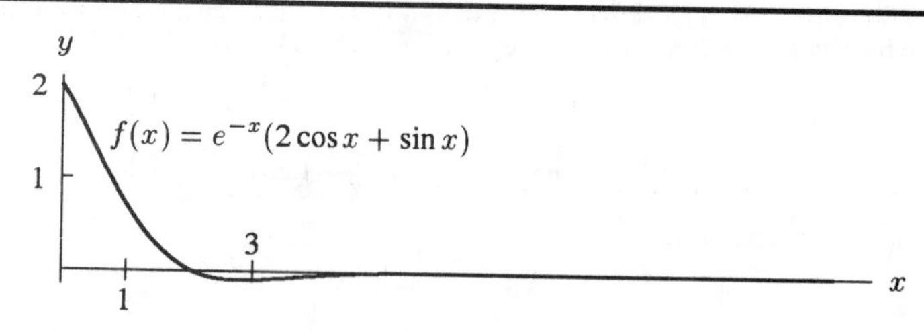

The maximum value of $f(x)$ is 2, which occurs when $x = 0$. The minimum appears to be $y \approx -0.94$, at $x \approx 2.82$.

5. (a) We start the time count on 1/1/90, so substituting $t = 0$ into $f(t)$ gives us the value of b, since both mt and $A \sin \frac{\pi t}{6}$ are equal to zero when $t = 0$. Thus, $b = f(0) = 20$. We see that in the 12 month period between 1/1/90 and 1/1/91 the value of the stock rose by \$30.00. Therefore, the linear component grows at the rate of \$30.00/year, or in terms of months, $\frac{30}{12} =$ \$2.50/month. So $m = 2.5$. Thus we have

$$P = f(t) = 2.5t + 20 + A\sin\frac{\pi t}{6}.$$

At an arbitrary data point, say $(4/1/90, 37.50)$, we can solve for A. Since January 1 corresponds to $t = 0$, April 1 is $t = 3$. We have

$$37.50 = f(3) = 2.5(3) + 20 + A\sin\frac{3\pi}{6} = 7.5 + 20 + A\sin\frac{\pi}{2} = 27.5 + A.$$

Simplifying gives $A = 10$, and the function is

$$f(t) = 2.5t + 20 + 10\sin\frac{\pi t}{6}.$$

(b) The stock appreciates the most during the months when the sine function climbs the fastest. By looking at Figure 7.15 we see that this occurs roughly when $t = 0$ and $t = 11$, January and December.

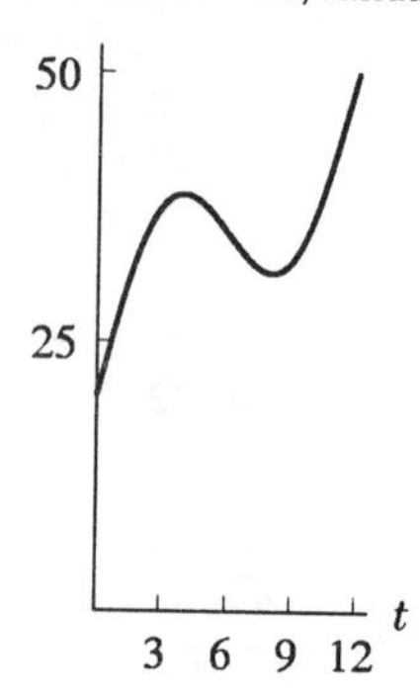

Figure 7.15

(c) Again, we look to Figure 7.15 to see when the graph actually decreases. It seems that the graph is decreasing roughly between the fourth and eighth months, that is, between May and September.

9. (a) Types of video games are trendy for a length of time, during which they are extremely popular and sales are high, later followed by a cooling down period as the users become tired of that particular game type. The game players then become interested in a different game type — and so on.

(b) The sales graph does not fit the shape of the sine or cosine curve, and we would have to say that neither of those functions would give us a reasonable model. However from 1979–1989 the graph does have a basic negative cosine shape, but the amplitude varies.

(c) One way to modify the amplitude over time is to multiply the sine (or cosine) function by an exponential function, such as e^{kt}. So we choose a model of the form

$$s(t) = e^{kt}(-a\cos(Ct) + D),$$

where t is the number of years since 1979. Note the $-a$, which is due to the graph looking like an inverted cosine at 1979. The average value starts at about 1.6, and the period appears to be about 6 years. The amplitude is initially about 1.4, which is the distance between the average value of 1.6 and the first peak value of 3.0. This means

$$s(t) = e^{kt}\left(-1.4\cos\left(\frac{2\pi}{6}t\right) + 1.6\right).$$

By trial and error on your graphing calculator, you can arrive at a value for the parameter k. A reasonable choice is $k = 0.05$, which gives

$$s(t) = e^{0.05t}\left(-1.4\cos\left(\frac{2\pi}{6}t\right) + 1.6\right).$$

(d)

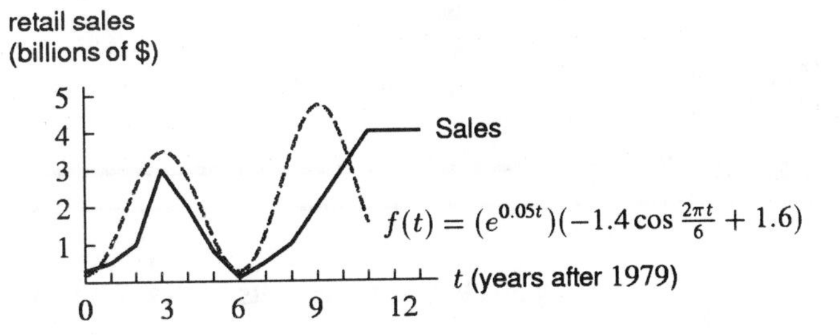

Notice that even though multiplying by the exponential function does increase the amplitude over time, it does not increase the period. Therefore, our model $s(t)$ does not fit the actual curve all that well.

(e) The predicted 1993 sales volume is $f(14) = 4.6$ billion dollars.

Solutions for Section 7.6

1. Figure 7.16 shows that at 12 noon, we have:
In Cartesian coordinates, $H = (0, 3)$. In polar coordinates, $H = (3, \pi/2)$; that is $r = 3, \theta = \pi/2$. In Cartesian coordinates, $M = (0, 4)$. In polars coordinates, $M = (4, \pi/2)$, that is $r = 4, \theta = \pi/2$.

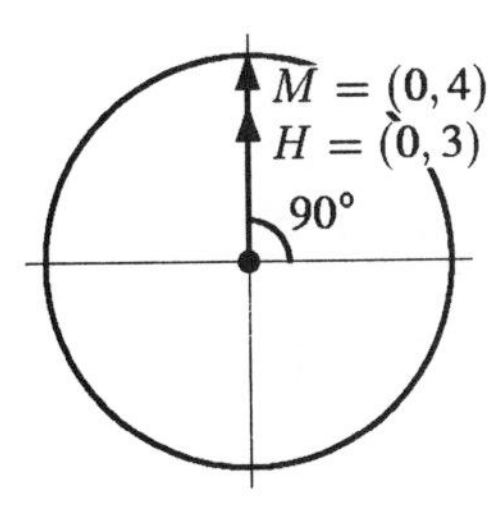

Figure 7.16

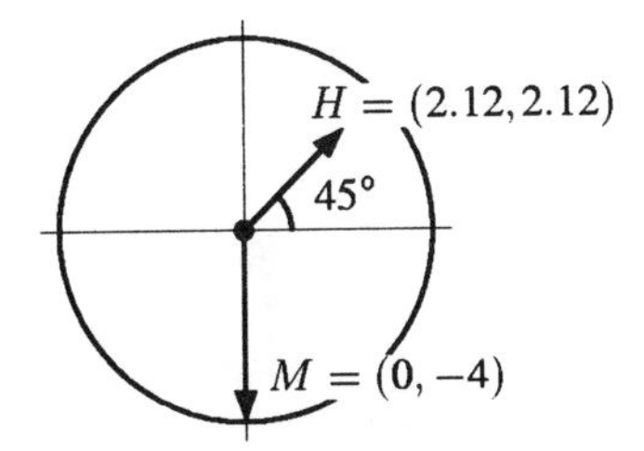

Figure 7.17

5. Figure 7.17 shows that at 1:30 pm, the polar coordinates of the point H (halfway between 1 and 2 on the clock face) are $r = 3$ and $\theta = 45° = \pi/4$. Thus, the Cartesian coordinates of H are given by

$$x = 3\cos\left(\frac{\pi}{4}\right) = \frac{3\sqrt{2}}{2} \approx 2.12, \quad y = 3\sin\left(\frac{\pi}{4}\right) = \frac{3\sqrt{2}}{2} \approx 2.12.$$

In Cartesian coordinates, $H \approx (2.12, 2.12)$. In polar coordinates, $H = (3, \pi/4)$. In Cartesian coordinates, $M = (0, -4)$. In polar coordinates, $M = (4, 3\pi/2)$.

9. The region is given by $\sqrt{8} \le r \le \sqrt{18}$ and $\pi/4 \le \theta \le \pi/2$.

13. There will be n loops. See Figures 7.18-7.21.

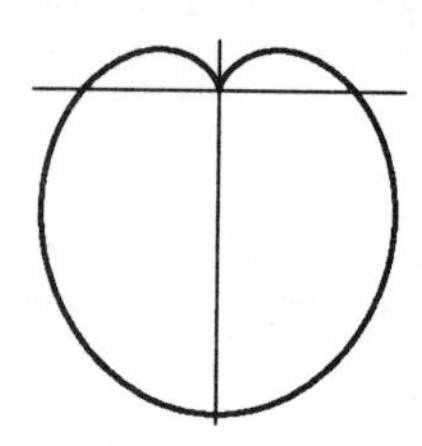

Figure 7.18: $n = 1$

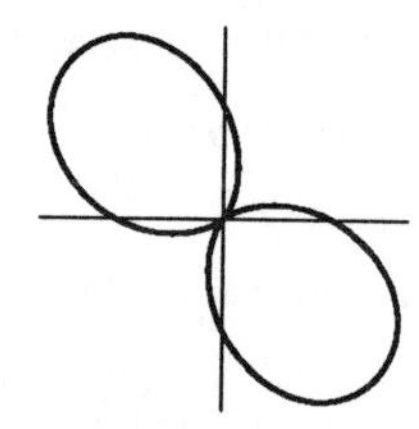

Figure 7.19: $n = 2$

Figure 7.20: $n = 3$

Figure 7.21: $n = 4$

17. Let $0 \leq \theta \leq 2\pi$ and $3/16 \leq r \leq 1/2$.

Solutions for Chapter 7 Review

1. We know all three sides of this triangle, but only one of its angles. We find the value of $\sin\theta$ and $\sin\phi$ in this right triangle:

$$\sin\theta = \frac{\text{opposite}}{\text{hypotenuse}} = \frac{3}{5} = 0.6$$

and

$$\sin\phi = \frac{\text{opposite}}{\text{hypotenuse}} = \frac{4}{5} = 0.8.$$

Using inverse sines, we know that if $\sin\phi = 0.8$, then $\phi = \sin^{-1}(0.8) \approx 53.1°$. Similarly $\sin\theta = 0.6$ means $\theta = \sin^{-1}(0.6) \approx 36.9°$. Notice $\phi + \theta = 90°$, which has to be true in a right triangle.

5. See Figure 7.22. By the Pythagorean theorem, $x = \sqrt{13^2 - 12^2} = 5$.

$$\sin\theta = \frac{12}{13}$$
$$\theta = \sin^{-1}\left(\frac{12}{13}\right)$$
$$\theta \approx 67.38°.$$

Thus $\varphi = 90° - \theta \approx 22.62°$.

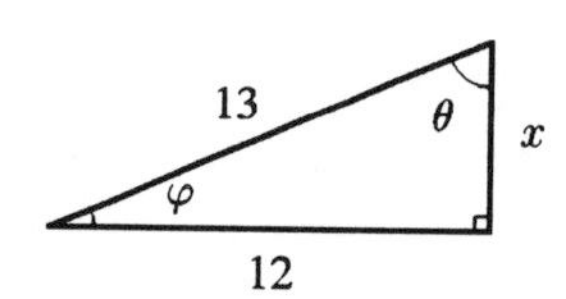

Figure 7.22

9. We have $2/7 = \cos 2\theta = 2\cos^2\theta - 1$. Solving for $\cos\theta$ gives

$$2\cos^2\theta = \frac{9}{7}$$
$$\cos^2\theta = \frac{9}{14}$$

Since θ is in the first quadrant, $\cos\theta = +\sqrt{9/14} = 3/\sqrt{14}$.

13.

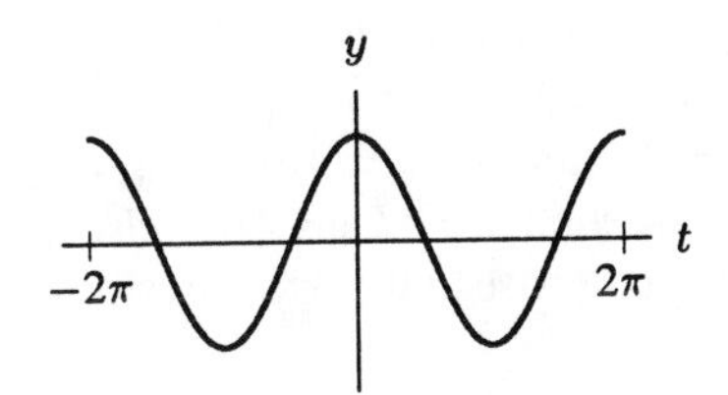

Figure 7.23: Graphs showing $\cos(t) = \sin\left(t + \frac{\pi}{2}\right)$

They appear to be the same graph. This suggests the truth of the identity $\cos t = \sin(t + \frac{\pi}{2})$.

17. Answers vary. If they use a 45° angle then they would measure an equal distance horizontally and vertically. See Figure 7.24.

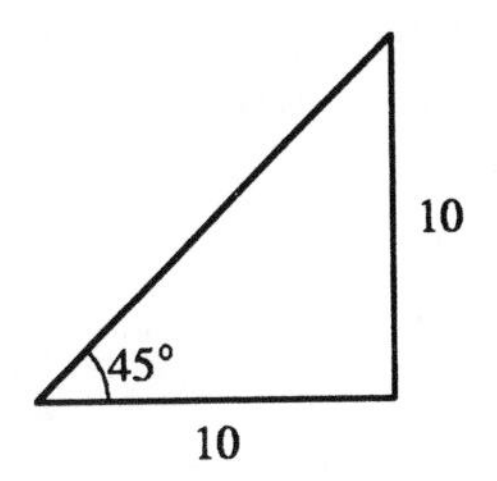

Figure 7.24

21. The graphs are found below.

(a)

(b)

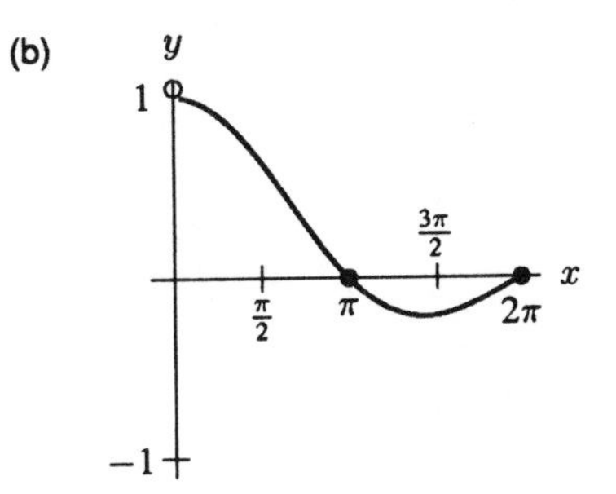

(c)

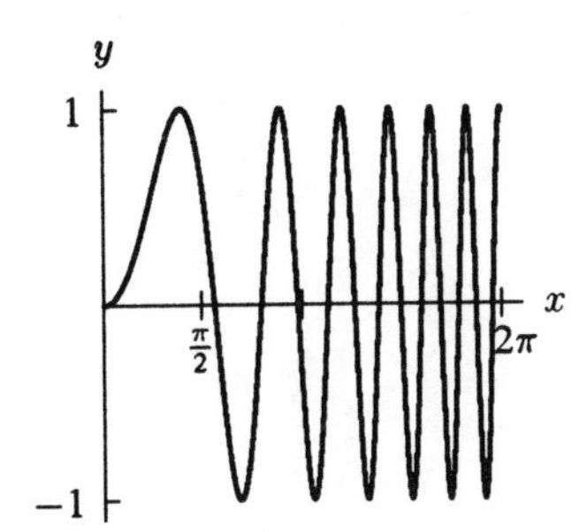

(d)

(e)

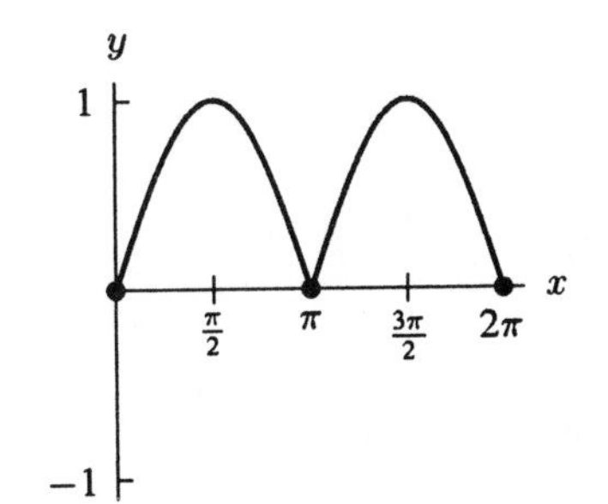

(f)

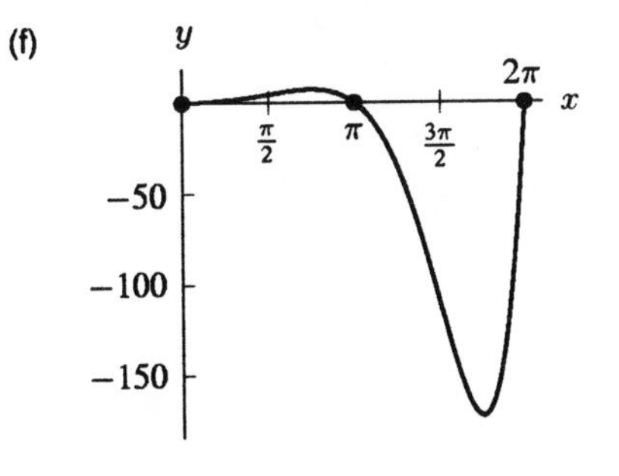

25. (a) Since B is the point $(1, 0)$, the circle has radius 1. Thus, $\sin\theta = \text{OE}$.
(b) From the definition of $\cos\theta$, we have $\cos\theta = \text{OA}$.
(c) In ΔODB, we have $OB = 1$. Since $\tan\theta = \text{Opp}/\text{Adj} = \text{DB}/1$, we have $\tan\theta = \text{DB}$.
(d) Let P be the point of intersection of $\overline{FC}$ and $\overline{OD}$. Use the fact that ΔOFP and ΔOEP are similar, and form a proportion of the hypotenuse to the one unit side of the larger triangle.

$$\frac{\text{OF}}{1} = \frac{1}{\text{OE}} \quad \text{or} \quad \text{OF} = \frac{1}{\sin\theta}$$

(e) Use the fact that ΔOCP and ΔOAP are similar, and write

$$\frac{\text{OC}}{1} = \frac{1}{\text{OA}} \quad \text{or} \quad \text{OC} = \frac{1}{\cos\theta}$$

(f) Use the fact that ΔGOH and ΔDOB are similar, and write

$$\frac{\text{GH}}{1} = \frac{1}{\text{DB}} \quad \text{or} \quad \text{GH} = \frac{1}{\tan\theta}$$

CHAPTER EIGHT

Solutions for Section 8.1

1. To construct a table of values for r, we must evaluate $r(0), r(1), \ldots, r(5)$. Starting with $r(0)$, we have
$$r(0) = p(q(0)).$$
Therefore
$$r(0) = p(5) \qquad \text{(because } q(0) = 5\text{)}$$
Using the table given in the problem, we have
$$r(0) = 4.$$
We can repeat this process for $r(1)$:
$$r(1) = p(q(1)) = p(2) = 5.$$
Similarly,
$$\begin{aligned} r(2) &= p(q(2)) = p(3) = 2 \\ r(3) &= p(q(3)) = p(1) = 0 \\ r(4) &= p(q(4)) = p(4) = 3 \\ r(5) &= p(q(5)) = p(8) = \text{ undefined.} \end{aligned}$$
These results have been compiled in Table 8.1.

TABLE 8.1

x	0	1	2	3	4	5
$r(x)$	4	5	2	0	3	–

5.

x	$f(x)$	$g(x)$	$h(x)$
0	1	2	5
1	9	0	1
2	5	1	9

According to the table, $h(0) = 5$. By definition, it is also true that $h(0) = f(g(0))$. Since $g(0) = 2$, $f(g(0)) = f(2)$. Put these pieces together:
$$\begin{aligned} h(0) &= 5 \\ h(0) &= f(g(0)) = f(2) \end{aligned}$$
So, $f(2) = 5$. We have $h(0) = 5 = f(g(0)) = f(2)$, so $f(2) = 5$. Also, $h(1) = f(g(1)) = f(0) = 1$. Finally, $h(2) = f(g(2)) = f(1) = 9$.

9. (a) $-3g(x) = -3(x^2 + x)$.
   (b) $g(1) - x = (1^2 + 1) - x = 2 - x$.
   (c) $g(x) + \pi = (x^2 + x) + \pi = x^2 + x + \pi$.
   (d) $\sqrt{g(x)} = \sqrt{x^2 + x}$.
   (e) $g(1)/(x + 1) = (1^2 + 1)/(x + 1) = 2/(x + 1)$.
   (f) $(g(x))^2 = (x^2 + x)^2$.

13. Substituting the expression $x^2 + 1$ for the x term in the formula for $h(x)$ gives $\sqrt{x^2 + 1}$.

17. $m(k(x)) = \dfrac{1}{k(x) - 1} = \dfrac{1}{x^2 - 1}$

21. Substituting $m(x) = 1/(x-1)$ into $n(x)$ gives

$$n(m(x)) = \frac{2(m(x))^2}{m(x)+1} = \frac{2\left(\frac{1}{x-1}\right)^2}{\frac{1}{x-1}+1} = \frac{2\cdot\frac{1}{(x-1)^2}}{\frac{1}{x-1}+\frac{x-1}{x-1}} = \frac{\frac{2}{(x-1)^2}}{\frac{1+x-1}{x-1}}$$

$$= \frac{2}{(x-1)^2}\cdot\frac{x-1}{x} = \frac{2}{x(x-1)}.$$

25. Since $n(x) = 2x^2/(x+1)$, we have

$$n(n(x)) = \frac{2(n(x))^2}{n(x)+1} = \frac{2(2x^2/(x+1))^2}{2x^2/(x+1)+1} = \frac{2\cdot 4x^2}{(x+1)^2}\cdot\frac{x+1}{2x^2+x+1}$$

$$= \frac{8x^4}{(x+1)(2x^2+x+1)}.$$

29. First we calculate

$$f(x+h) - f(x) = \frac{1}{x+h} - \frac{1}{x} = \frac{x}{x(x+h)} - \frac{x+h}{x(x+h)}$$

$$= \frac{x-(x+h)}{x(x+h)} = \frac{-h}{x(x+h)}$$

Then

$$\frac{f(x+h)-f(x)}{h} = \frac{\frac{-h}{x(x+h)}}{h} = \frac{-h}{x(x+h)}\cdot\frac{1}{h} = \frac{-1}{x(x+h)}$$

33. If $f(x) = u(v(x))$, then one solution is $u(x) = \sqrt{x}$ and $v(x) = 3 - 5x$.

37. One possible solution is $j(x) = u(v(x))$ where $u(x) = 1 - x$ and $v(x) = \sqrt{x}$

41. If $g(x) = u(v(x))$, then one solution is $u(x) = \frac{1}{x}$ and $v(x) = 1 - x$.

45. One possible solution is $K(x) = u(v(x))$ where $u(x) = \sqrt{x}$ and $v(x) = 1 - 4x^2$.

49. These are possible decompositions. There could be others.
$o(x) = u(v(w(x)))$ where $u(x) = 1 - x$, $v(x) = \sqrt{x}$ and $w(x) = x - 1$.

53. $h(x) = x^3$

57. $g(f(x)) = (f(x))^2 + 3 = \underbrace{(x+1)^2 + 3}_{\text{This was given}}$. Thus one possibility is $f(x) = x + 1$.

Since $(x+1)^2 = (-(x+1))^2$, another possibility is $f(x) = -(x+1)$. There are others.

61. If $s(x) = 5 + \frac{1}{x+5} + x = x + 5 + \frac{1}{x+5}$ and $k(x) = x + 5$, then

$$s(x) = k(x) + \frac{1}{k(x)}.$$

However, $s(x) = v(k(x))$, so

$$v(k(x)) = k(x) + \frac{1}{k(x)}.$$

This is possible if

$$v(x) = x + \frac{1}{x}.$$

65. (a) The function f represents the exchange is dollars to yen, and 1 dollar buys 84.62 yen. Thus, each of the x dollars buys 84.62 yen, for a total of $84.62x$ yen. Therefore,

$$f(x) = 84.62x.$$

Referring to the table, we see that 1 dollar purchases 1.0908 European Union euros. If x dollars are invested, each of the x dollars will buy 1.0908 euros, for a total of $1.0908x$ euros. Thus,

$$g(x) = 1.0908x.$$

Finally, we see from the table that 1 yen buys 0.0129 euros. Each of the x yen invested buys 0.0129 euros, so $0.0129x$ euros can be purchased. Therefore,

$$h(x) = 0.0129x.$$

(b) We evaluate $h(f(1000))$ algebraically. Since $f(1000) = 84.62(1000) = 84620$, we have

$$\begin{aligned} h(f(1000)) &= h(84620) \\ &= 0.0129(84620) \\ &= 1091.598. \end{aligned}$$

To interpret this statement, we break the problem into steps. First, we see that $f(1000) = 84{,}620$ means 1000 dollars buy 84,620 yen. Second, we see that $h(84620) = 1091.598$ means that 84620 yen buys 1091.598 euros. In other words, $h(f(1000)) = 1091.598$ represents a trade of \$1000 for 84,620 yen which is subsequently traded for 1091.598 euros (Of course, a direct trade of \$1000 would yield 1090.8 euros).

Solutions for Section 8.2

1. (a) We have

$$R = 150 + 5T.$$

Solving for T we have

$$5T = R - 150,$$

so

$$T = f^{-1}(R) = \frac{1}{5}R - 30.$$

(b) Table 8.2 shows values of f, and Table 8.3 shows values of f^{-1}. Notice that the columns in the two tables are reversed. This is because input values of f are output values of f^{-1}, and input values of f^{-1} are output values of f.

TABLE 8.2 *Values of f*

T, temperature (°C)	$R = f(T)$, resistance (ohms)
−20	50
−10	100
0	150
10	200
20	250
30	300
40	350
50	400

TABLE 8.3 *Values of f^{-1}*

R, resistance (ohms)	$T = f^{-1}(R)$, temperature (°C)
50	−20
100	−10
150	0
200	10
250	20
300	30
350	40
400	50

5. It is invertible.

9. We let $y = f(x) = x^3$ and find a formula for $x = f^{-1}(y)$. Since

$$y = x^3,$$

solving for x gives

$$x = y^{1/3},$$

so $f^{-1}(y) = y^{1/3}$. (It should come as no surprise that the inverse of the cubing function is the cube root function.) In order to graph both functions on the same axes, we write the function f^{-1} using the same variables as the function f, thus:

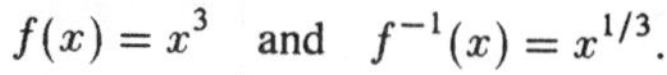

$$f(x) = x^3 \quad \text{and} \quad f^{-1}(x) = x^{1/3}.$$

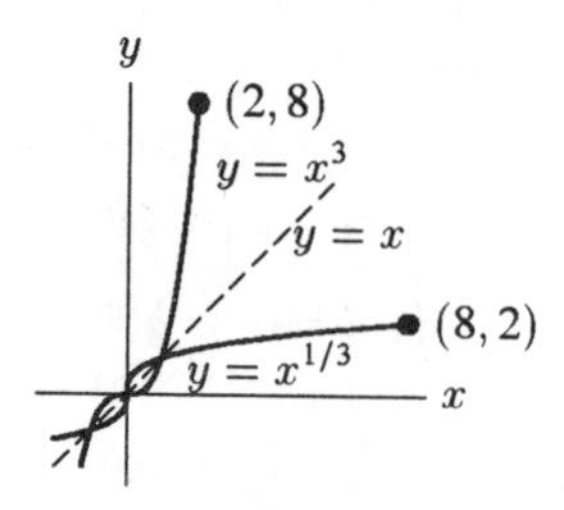

Figure 8.1: The graphs of $y = x^3$ and $y = x^{1/3}$. The graph and its inverse are symmetrical across the line $y = x$

Figure 8.1 shows the graphs of f and f^{-1}. Notice that these graphs are symmetric about the line $y = x$. For instance,

$$f(2) = 2^3 = 8 \quad \text{and} \quad f^{-1}(8) = 8^{1/3} = 2,$$

hence the point $(2, 8)$ lies on the graph of f and the point $(8, 2)$ lies on the graph of f^{-1}.

13. Reading the values from the graph, we get:

$$f(0) = 1.5, \quad f^{-1}(0) = 2.5, \quad f(3) = -0.5, \quad f^{-1}(3) = -5.$$

Ranking them in order from least to greatest, we get:

$$f^{-1}(3) < f(3) < 0 < f(0) < f^{-1}(0) < 3.$$

17. One way to check that these functions are inverses is to make sure they satisfy the identities $f(f^{-1}(x)) = x$ and $f^{-1}(f(x)) = x$.

$$\begin{aligned} f(f^{-1}(x)) &= 1 + 7\left(\sqrt[3]{\frac{x-1}{7}}\right)^3 \\ &= 1 + 7\left(\frac{x-1}{7}\right) \\ &= 1 + (x - 1) = x. \end{aligned}$$

Also,

$$\begin{aligned} f^{-1}(f(x)) &= \sqrt[3]{\frac{1 + 7x^3 - 1}{7}} \\ &= \sqrt[3]{x^3} = x. \end{aligned}$$

Thus, $f^{-1}(x) = \sqrt[3]{\dfrac{x-1}{7}}$.

21. Solve for x in $y = h(x) = 12x^3$:

$$\begin{aligned} y &= 12x^3 \\ x^3 &= \frac{y}{12} \\ x = h^{-1}(y) &= \sqrt[3]{\frac{y}{12}}. \end{aligned}$$

Writing h^{-1} in terms of x gives $h^{-1}(x) = \sqrt[3]{\dfrac{x}{12}}$.

25. Start with $x = f(f^{-1}(x))$ and substitute $y = f^{-1}(x)$. We have

$$\begin{aligned} x &= f(y) \\ x &= 10^y \\ \log(x) &= \log 10^y \\ \log(x) &= y \end{aligned}$$

Thus, $y = f^{-1}(x) = \log(x)$.

29. Start with $x = n(n^{-1}(x))$ and substitute $y = n^{-1}(x)$. We have

$$\begin{aligned} x &= n(y) \\ x &= \log(y-3) \\ 10^x &= 10^{\log(y-3)} \\ 10^x &= y-3 \\ y &= 10^x + 3 \end{aligned}$$

So $y = n^{-1}(x) = 10^x + 3$.

33. Start with $x = f(f^{-1}(x))$ and let $y = f^{-1}(x)$. Then $x = f(y)$ means

$$\begin{aligned} x &= \frac{3+2y}{2-5y} \\ x(2-5y) &= 3+2y \\ 2x - 5xy &= 3+2y \\ 2x - 3 &= 5xy + 2y \\ 2x - 3 &= y(5x+2) \\ y &= \frac{2x-3}{5x+2}, \end{aligned}$$

so $y = f^{-1}(x) = \dfrac{2x-3}{5x+2}$.

37. Solve for x in $y = m(x) = \sqrt{(x+1)/x}$:

$$\begin{aligned} y &= \sqrt{\frac{x+1}{x}} \\ y^2 &= \frac{x+1}{x} \\ xy^2 &= x+1 \\ yx^2 - x &= 1 \\ x(y^2-1) &= 1 \\ x &= m^{-1}(y) = \frac{1}{y^2-1}. \end{aligned}$$

Writing m^{-1} in terms of x gives $m^{-1}(x) = \dfrac{1}{x^2-1}$. Since $y = m(x) > 1$ when $x > 0$, we take $y > 1$. Thus, $x > 1$ for $m^{-1}(x)$.

41. (a) To find the formula for an inverse function from the formula for the original function, solve for the independent variable. For example, if $y = f(x) = x^3$ is the original function, the inverse function is $x = f^{-1}(y) = y^{1/3}$, which is often written as $y = f^{-1}(x) = x^{1/3}$.

(b) Reflect the graph over the line $y = x$. For example, the graph of $f(x) = x^3$ can be flipped over the line $y = x$ to give the graph of $f^{-1}(x) = x^{1/3}$.

(c) As we saw in part (b), the graph of a relation is flipped about the line $y = x$ to get the graph of the inverse. $y = x$ is the line of symmetry. The example given in (b) shows this property.

45. We raise each side to the $1/(1.05)$ power:

$$x^{1.05} = 1.09$$
$$x = 1.09^{1/1.05}.$$

49. (a) $A = \pi r^2$
(b) The graph of the function in part (a) is in Figure 8.2.

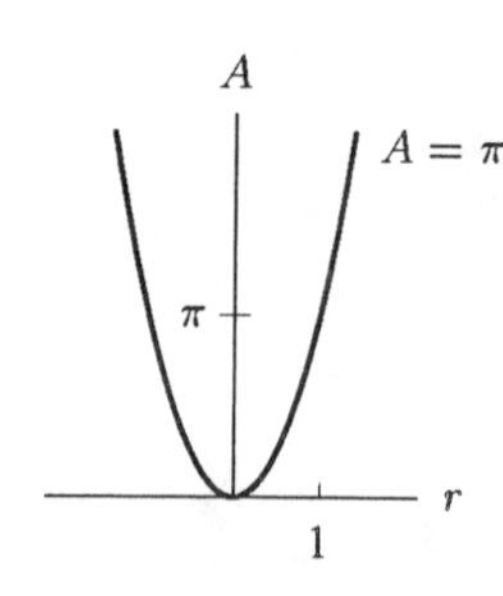

Figure 8.2

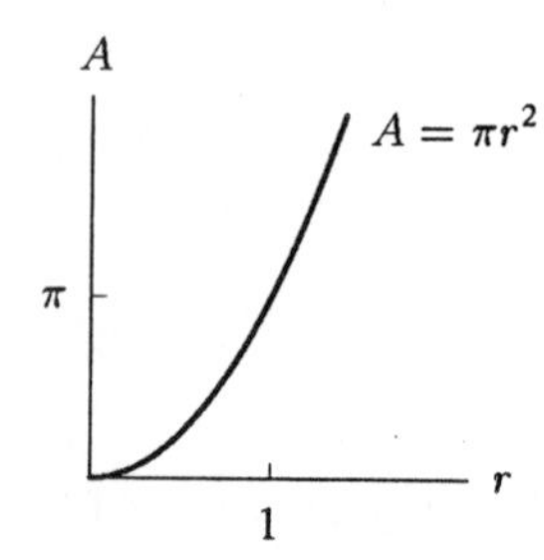

Figure 8.3

(c) Because a circle cannot have a negative radius, the domain is $r \geq 0$. See Figure 8.3.
(d) Solve the formula $A = f(r) = \pi r^2$ for r in terms of A:

$$r^2 = \frac{A}{\pi}$$
$$r = \pm\sqrt{\frac{A}{\pi}}$$

The range of the inverse function is the same as the domain of f, namely non-negative real numbers. Thus, we choose the positive root, and $f^{-1}(A) = \sqrt{\frac{A}{\pi}}$.
(e) We rewrite both functions to be y in terms of x, and graph. See Figure 8.4.

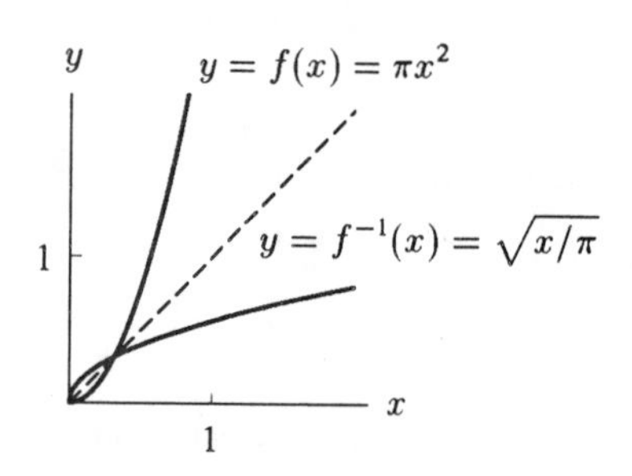

Figure 8.4

(f) Yes. If the function $A = \pi r^2$ refers to radius and area, its domain must be $r \geq 0$. On this domain the function is invertible, so radius is also a function of area.

53. (a) When $t = 0$, $P = 37.8(1.044)^0 = 37.8(1) = 37.8$. This tells us that the population of the town when $t = 0$ is 37,800. The growth factor, 1.044, tells us that the population is 104.4% of what it had been the previous year, or that the town grows by 4.4% each year.

(b) Since $f(t) = 37.8(1.044)^t$, then $f(50) = 37.8(1.044)^{50} \approx 325.5$ This tells us that there will be approximately 325,500 people after 50 years.

(c) To find $f^{-1}(P)$, which is the inverse function of $f(t)$, we need to solve

$$P = 37.8(1.044)^t$$

for t. Begin by dividing both sides by 37.8:

$$\frac{P}{37.8} = 1.044^t$$

Then, take the log of both sides, using the property $\log a^b = b \cdot \log a$.

$$\log\left(\frac{P}{37.8}\right) = \log 1.044^t = t \log 1.044.$$

So solving for t,

$$t = \frac{\log\left(\frac{P}{37.8}\right)}{\log 1.044}.$$

We can make this formula look a little simpler by recalling that $\log \frac{a}{b} = \log a - \log b$. The formula for our inverse function is now:

$$t = f^{-1}(P) = \frac{\log P - \log 37.8}{\log 1.044}.$$

(d) $f^{-1}(50) = \frac{\log 50 - \log 37.8}{\log 1.044} \approx 6.5$. It will take about 6.5 years for P to reach 50,000 people.

57. (a) $C(0)$ is the concentration of alcohol in the 100 ml solution after 0 ml of alcohol is removed. Thus, $C(0) = 99\%$.

(b) Note that there are initially 99 ml of alcohol and 1 ml of water.

$$C(x) = \begin{array}{c}\text{Concentration of alcohol}\\ \text{after removing } x \text{ ml}\end{array} = \frac{\text{Amount of alcohol remaining}}{\text{Amount of solution remaining}}$$

$$= \frac{\begin{array}{c}\text{Original amount}\\ \text{of alcohol}\end{array} - \begin{array}{c}\text{Amount of}\\ \text{alcohol removed}\end{array}}{\begin{array}{c}\text{Original amount}\\ \text{of solution}\end{array} - \begin{array}{c}\text{Amount of alcohol}\\ \text{removed}\end{array}} = \frac{99 - x}{100 - x}.$$

(c) If $y = C(x)$, then $x = C^{-1}(y)$. We have

$$\begin{aligned} y &= \frac{99 - x}{100 - x} \\ y(100 - x) &= 99 - x \\ 100y - xy &= 99 - x \\ x - xy &= 99 - 100y \\ x(1 - y) &= 99 - 100y \\ x &= \frac{99 - 100y}{1 - y}. \end{aligned}$$

Thus, $C^{-1}(y) = \dfrac{99 - 100y}{1 - y}$.

(d) The function $C^{-1}(y)$ tells us how much alcohol we need to remove in order to obtain a solution whose concentration is y.

61. Let $y = f(x)$. In order to find f^{-1}, we need to solve for x. But

$$y = g(h(x)), \text{ so } g^{-1}(y) = g^{-1}(g(h(x))) = h(x).$$

Moreover,

$$h^{-1}(g^{-1}(y)) = h^{-1}(h(x)) = x,$$

hence $x = f^{-1}(y) = h^{-1}(g^{-1}(y))$. So $f^{-1}(x) = h^{-1}(g^{-1}(x))$.

Solutions for Section 8.3

1. Since $h(x) = f(x) + g(x)$, we know that $h(-1) = f(-1) + g(-1) = -4 + 4 = 0$. Similarly, $j(x) = 2f(x)$ tells us that $j(-1) = 2f(-1) = 2(-4) = -8$. Repeat this process for each entry in the table.

TABLE 8.4

x	$h(x)$	$j(x)$	$k(x)$	$m(x)$
-1	0	-8	16	-1
0	0	-2	1	-1
1	2	4	0	0
2	6	10	1	0.2
3	12	16	16	0.5
4	20	22	81	9/11

5. (a) A formula for $h(x)$ would be

$$h(x) = f(x) + g(x).$$

To evaluate $h(x)$ for $x = 3$, we use this equation:

$$h(3) = f(3) + g(3).$$

Since $f(x) = x + 1$, we know that

$$f(3) = 3 + 1 = 4.$$

Likewise, since $g(x) = x^2 - 1$, we know that

$$g(3) = 3^2 - 1 = 9 - 1 = 8.$$

Thus, we have

$$h(3) = 4 + 8 = 12.$$

To find a formula for $h(x)$ in terms of x, we substitute our formulas for $f(x)$ and $g(x)$ into the equation $h(x) = f(x) + g(x)$:

$$h(x) = \underbrace{f(x)}_{x+1} + \underbrace{g(x)}_{x^2-1}$$

$$h(x) = x + 1 + x^2 - 1 = x^2 + x.$$

To check this formula, we use it to evaluate $h(3)$, and see if it gives $h(3) = 12$, which is what we got before. The formula is $h(x) = x^2 + x$, so it gives

$$h(3) = 3^2 + 3 = 9 + 3 = 12.$$

This is the result that we expected.

(b) A formula for $j(x)$ would be

$$j(x) = g(x) - 2f(x).$$

To evaluate $j(x)$ for $x = 3$, we use this equation:

$$j(3) = g(3) - 2f(3).$$

We already know that $g(3) = 8$ and $f(3) = 4$. Thus,

$$j(3) = 8 - 2 \cdot 4 = 8 - 8 = 0.$$

To find a formula for $j(x)$ in terms of x, we again use the formulas for $f(x)$ and $g(x)$:

$$\begin{aligned} j(x) &= \underbrace{g(x)}_{x^2-1} - 2\underbrace{f(x)}_{x+1} \\ &= (x^2 - 1) - 2(x + 1) \\ &= x^2 - 1 - 2x - 2 \\ &= x^2 - 2x - 3. \end{aligned}$$

We check this formula using the fact that we already know $j(3) = 0$. Since we have $j(x) = x^2 - 2x - 3$,

$$j(3) = 3^2 - 2 \cdot 3 - 3 = 9 - 6 - 3 = 0.$$

This is the result that we expected.

(c) A formula for $k(x)$ would be

$$k(x) = f(x)g(x).$$

Evaluating $k(3)$, we have

$$k(3) = f(3)g(3) = 4 \cdot 8 = 32.$$

A formula in terms of x for $k(x)$ would be

$$\begin{aligned} k(x) &= \underbrace{f(x)}_{x+1} \cdot \underbrace{g(x)}_{x^2-1} \\ &= (x+1)(x^2-1) \\ &= x^3 - x + x^2 - 1 \\ &= x^3 + x^2 - x - 1. \end{aligned}$$

To check this formula,

$$k(3) = 3^3 + 3^2 - 3 - 1 = 27 + 9 - 3 - 1 = 32,$$

which agrees with what we already knew.

(d) A formula for $m(x)$ would be

$$m(x) = \frac{g(x)}{f(x)}.$$

Using this formula, we have

$$m(3) = \frac{g(3)}{f(3)} = \frac{8}{4} = 2.$$

To find a formula for $m(x)$ in terms of x, we write

$$\begin{aligned} m(x) = \frac{g(x)}{f(x)} &= \frac{x^2-1}{x+1} \\ &= \frac{(x+1)(x-1)}{(x+1)} \\ &= x - 1 \text{for } x \neq -1 \end{aligned}$$

We were able to simplify this formula by first factoring the numerator of the fraction $\frac{x^2-1}{x+1}$. To check this formula,

$$m(3) = 3 - 1 = 2,$$

which is what we were expecting.

(e) We have

$$n(x) = \left(f(x)\right)^2 - g(x).$$

This means that

$$\begin{aligned} n(3) &= \left(f(3)\right)^2 - g(3) \\ &= (4)^2 - 8 \\ &= 16 - 8 \\ &= 8. \end{aligned}$$

A formula for $n(x)$ in terms of x would be

$$\begin{aligned} n(x) &= (f(x))^2 - g(x) \\ &= (x+1)^2 - (x^2-1) \\ &= x^2 + 2x + 1 - x^2 + 1 \\ &= 2x + 2. \end{aligned}$$

To check this formula,

$$n(3) = 2 \cdot 3 + 2 = 8,$$

which is what we were expecting.

9. $j(x) = \frac{2x-1}{\frac{1}{x}} = x \cdot (2x-1) = 2x^2 - x.$

13. (a) Since $P(t)$ is an exponential function we know that its formula will be of the form

$$P(t) = P_0 \cdot A^t$$

where P_0 is the initial population and A is the rate at which the population changes. In the table, we are given two pairs of values for the function $P(t)$. Thus, we know

$$4{,}500{,}000{,}000 = P(5) = P_0 \cdot A^5$$

and

$$5{,}695{,}300{,}000 = P(15) = P_0 \cdot A^{15}.$$

Dividing the second equation by the first we get

$$\frac{5{,}695{,}300{,}000}{4{,}500{,}000{,}000} = \frac{P_0 \cdot A^{15}}{P_0 \cdot A^5}.$$

That is

$$1.2656 \approx \frac{A^{15}}{A^5} = A^{15-5} = A^{10}.$$

Solving for A we get

$$A = \sqrt[10]{1.2656} \approx 1.0238.$$

Thus we have

$$P(t) = P_0 \cdot (1.0238)^t.$$

To solve for P_0 we plug in the first pair of values and get

$$4{,}500{,}000{,}000 = P_0 \cdot (1.0238)^5 \approx P_0 \cdot 1.125.$$

Thus we get

$$P_0 \approx 4{,}000{,}000{,}000.$$

Thus the formula for $P(t)$ is

$$P(t) = 4{,}000{,}000{,}000 \cdot (1.0238)^t.$$

Note: Due to round-off errors, if you substitute 15 for t, you will not get exactly 5,695,300,000 but the result is close enough.

(b) Looking at our formula for $P(t)$ we see that each year the population changes at a rate of 1.0238. That is, ever year the population increases by 2.38%.

(c) Since $N(t)$ is linear, it will be of the form

$$N(t) = mt + b$$

where m is the slope and b is the y-intercept. We know that the slope of $N(t)$ must be

$$m = \frac{30{,}000 - 21{,}000}{0 - 10} = -900.$$

Also, at $t = 0$ we have $N(t) = 30{,}000$. Thus

$$b = 30{,}000$$

and

$$N(t) = 30{,}000 - 900t.$$

(d) Plugging into our formulas for $N(t)$ and $P(t)$ we get

TABLE 8.5

t, time in years	$N(t)$, number of warheads	$P(t)$, world population
0	30,000	4,000,000,000
5	25,500	4,500,000,000
10	21,000	5,060,700,000
15	16,500	5,695,300,000

Taking the quotient of these values we get

TABLE 8.6

t, time in years	$f(t)$
0	0.00000750
5	0.00000567
10	0.00000415
15	0.00000290

(e) We know that

$$f(t) = \frac{\text{linear function}}{\text{exponential function}}.$$

Since a linear function is of the form

$$\text{linear function} = mt + b$$

and an exponential function is of the form

$$\text{exponential function} = P_0 \cdot A^t$$

we know that $f(t)$ will be of the form

$$f(t) = \frac{mt+b}{P_0 \cdot A^t}.$$

This function is neither linear nor exponential, as $m \neq 0$ in our case.

(f) The function $f(t)$ tells us, at a given time t, per capita number of warheads.

17. (a) Since the zeros of this quadratic function are 0 and 4, the formula for this function will be of the form $f(x) = kx(x-4)$, which is the product of two linear functions, $a(x) = kx$ and $b(x) = x - 4$. Figure 8.5 shows a possible graph of these two functions.

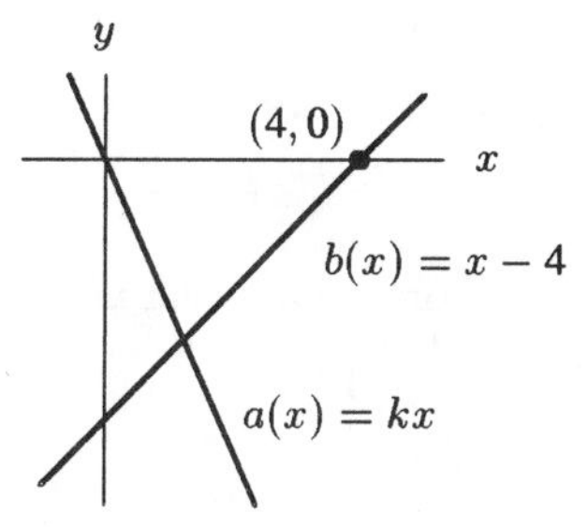

Figure 8.5

(b) If the formula of a quadratic function could be written as the product of two linear functions, $g(x) = (ax+b)(cx+d)$, then the function must have zeros at $-\frac{b}{a}$ and $-\frac{d}{c}$ (from $ax + b = 0$ and $cx + d = 0$). So any such quadratic function would have at least one zero (when $-\frac{b}{a}$ equals $-\frac{d}{c}$) and, more likely, two zeros. The function $y = q(x)$ has no zeros and therefore, cannot be the product of two linear functions.

21. The statement is false. For example, if $f(x) = x$ and $g(x) = x^2$, then $f(x) \cdot g(x) = x^3$. In this case, $f(x) \cdot g(x)$ is an odd function, but $g(x)$ is an even function.

25. In order to evaluate $h(3)$, we need to express the formula for $h(x)$ in terms of $f(x)$ and $g(x)$. Factoring gives
$$h(x) = C^{2x}(kx^2 + B + 1).$$
Since $g(x) = C^{2x}$ and $f(x) = kx^2 + B$, we can re-write the formula for $h(x)$ as
$$h(x) = g(x) \cdot (f(x) + 1).$$
Thus,
$$\begin{aligned} h(3) &= g(3) \cdot (f(3) + 1) \\ &= 5(7 + 1) \\ &= 40. \end{aligned}$$

Solutions for Chapter 8 Review

1. (a) $f(2x) = (2x)^2 + (2x) = 4x^2 + 2x$
   (b) $g(x^2) = 2x^2 - 3$
   (c) $h(1 - x) = \dfrac{(1 - x)}{1 - (1 - x)} = \dfrac{1 - x}{x}$
   (d) $(f(x))^2 = (x^2 + x)^2$
   (e) Since $g(g^{-1}(x)) = x$, we have
$$\begin{aligned} 2g^{-1}(x) - 3 &= x \\ 2g^{-1}(x) &= x + 3 \\ g^{-1}(x) &= \frac{x + 3}{2}. \end{aligned}$$
   (f) $(h(x))^{-1} = \left(\dfrac{x}{1 - x}\right)^{-1} = \dfrac{1 - x}{x}$
   (g) $f(x)g(x) = (x^2 + x)(2x - 3)$
   (h) $h(f(x)) = h(x^2 + x) = \dfrac{x^2 + x}{1 - (x^2 + x)} = \dfrac{x^2 + x}{1 - x^2 - x}$

5. (a) Since each of 7 people uses 2 gallons a day, altogether 14 gallons a day are used. After t days, $14t$ gallons have been used, so since there were originally 800 gallons,
$$\text{Fresh water remaining} = f(t) = 800 - 14t \text{ gallons.}$$
   (b) (i) $f(0) = 800$ gallons. This is the original amount of water brought to the island.
   (ii) This represents the time when they will run out of water. To find $f^{-1}(0)$ we solve:
$$\begin{aligned} 800 - 14t &= 0 \\ 14t &= 800 \\ t &= \frac{800}{14} \approx 57.1 \quad \text{days} \end{aligned}$$
   Since $f^{-1}(0) \approx 57.1$, they will run out of water after 57.1 days.
   (iii) Want to find t such that $f(t) = \frac{1}{2}f(0)$:
$$\begin{aligned} \frac{1}{2}f(0) &= 400 \\ 400 &= 800 - 14t \\ 14t &= 400 \\ t &\approx 28.6 \text{ days.} \end{aligned}$$
   This t value is the time when half of the original water is gone.

(iv) Substituting for $f(t)$ gives

$$800 - f(t) = 800 - (800 - 14t) = 14t.$$

This represents the total amount of water used in t days.

9. The graph is given in Figure 8.6.

13. The graph is given in Figure 8.7.

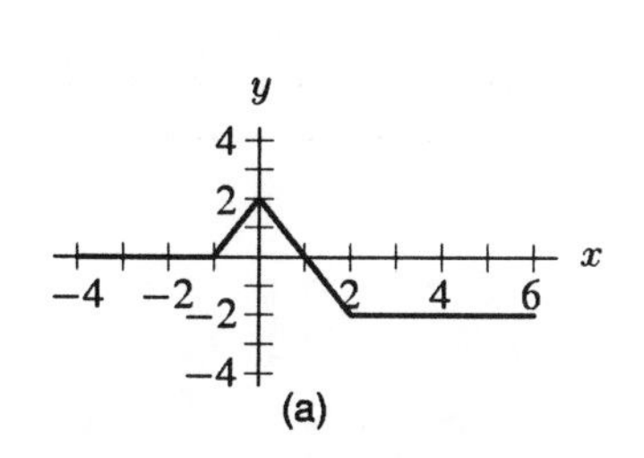

Figure 8.6

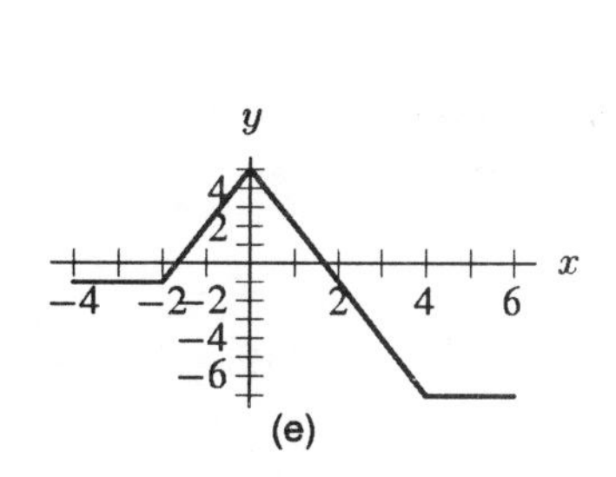

Figure 8.7

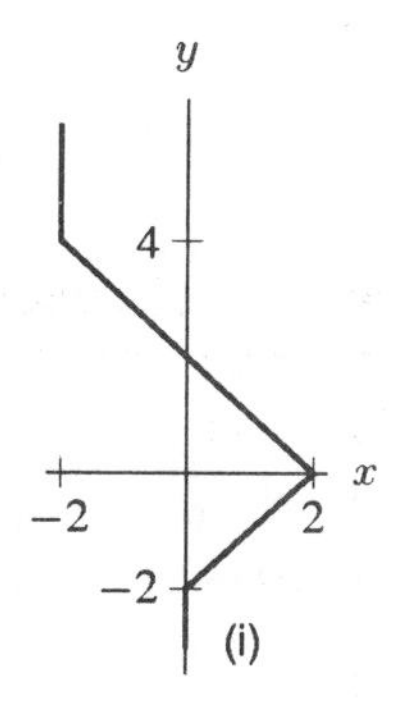

Figure 8.8

17. The graph is given in Figure 8.8.

21.

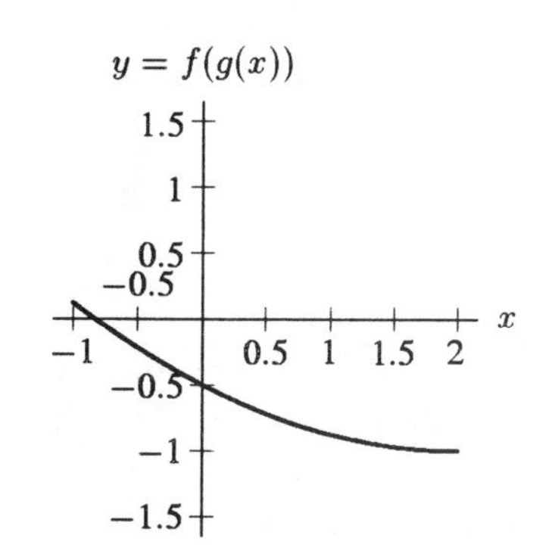

Figure 8.9

25. The graph is in Figure 8.10.

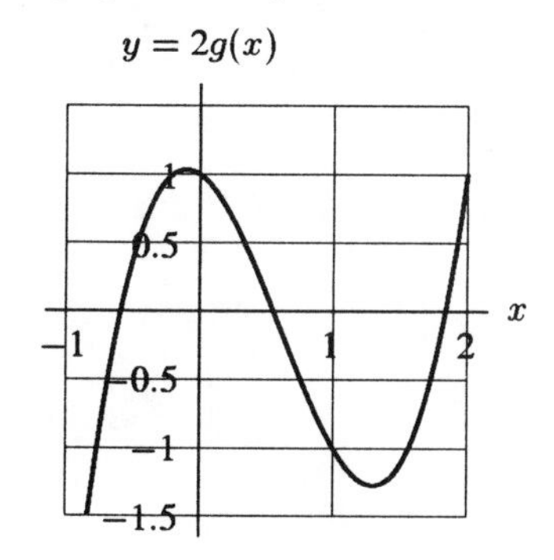

Figure 8.10

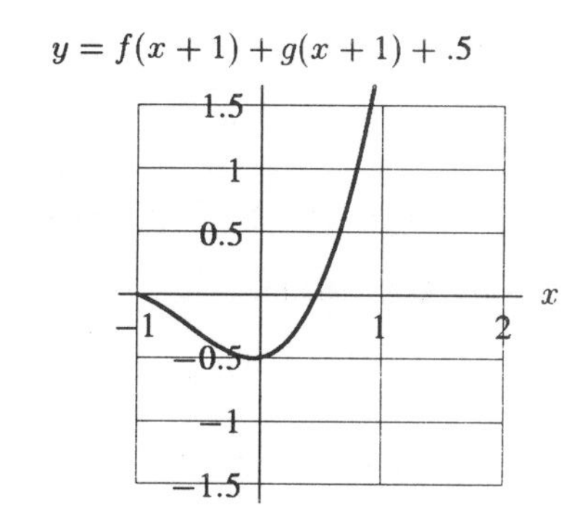

Figure 8.11

29. The graph is in Figure 8.11.

33. (a)

TABLE 8.7

Hours worked	Hours of study time
$0 \le h < 4$	39
$4 \le h < 8$	31
$8 \le h < 12$	25
$12 \le h < 16$	20
$16 \le h < 20$	16
$20 \le h < 24$	14
$24 \le h < 28$	12

(b)

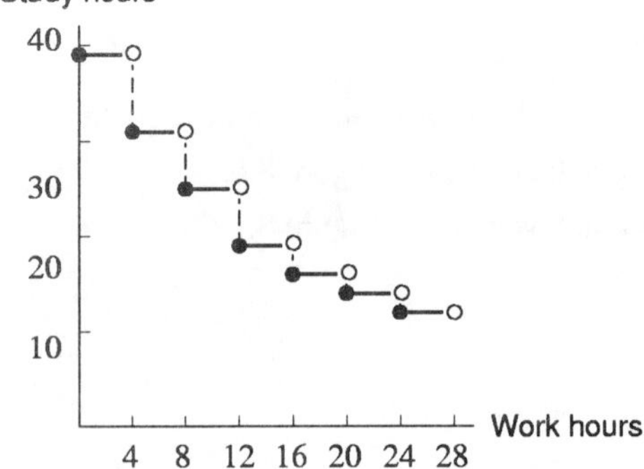

This relationship is not linear.

(c) There are many situations: for example, 21 hours of job time reduced to 13 hours saves 8 hours, but results in an increase of 6 hours in study time leaving 2 more hours of leisure time.

37. $g(q) < g(p)$ because the cost per square foot of building office space decreases as the total square footage increases. $g(p) < f(p)$ since the total cost of building more than one square foot is greater than the cost per square foot. $f(p) < f(q)$ since the total cost of building office space increases as the square footage increases. So

$$g(q) < g(p) < f(p) < f(q).$$

41. $g(x) = x + 1$ and $h(x) = 2x$

45. $g(x) = \dfrac{1}{x^2}$ and $h(x) = x + 4$

49. Since $h(g(x)) = h(2x + 5)$, we have

$$h(2x+5) = \frac{2x+5}{1+\sqrt{2x+5}},$$

which means

$$h(x) = \frac{x}{1+\sqrt{x}}.$$

53. Start with $x = j(j^{-1}(x))$ and substitute $y = j^{-1}(x)$. We have

$$\begin{aligned} x &= j(y) \\ x &= \sqrt{1+\sqrt{y}} \\ x^2 &= 1+\sqrt{y} \\ x^2 - 1 &= \sqrt{y} \\ (x^2-1)^2 &= y \end{aligned}$$

Therefore,

$$j^{-1}(x) = (x^2-1)^2.$$

57. Start with $x = f(f^{-1}(x))$ and substitute $y = f^{-1}(x)$. We have

$$\begin{aligned}
x &= f(y)\\
x &= \frac{3\cdot 2^y+1}{3\cdot 2^y+3}\\
(3\cdot 2^y+3)\cdot x &= 3\cdot 2^y+1\\
3x(2^y)+3x &= 3\cdot 2^y+1\\
3x(2^y)-3\cdot 2^y &= 1-3x\\
2^y(3x-3) &= 1-3x\\
2^y &= \frac{1-3x}{3x-3}\\
\log 2^y &= \log\frac{1-3x}{3x-3}\\
y\log 2 &= \log\frac{1-3x}{3x-3}\\
y &= \frac{\log\left(\frac{1-3x}{3x-3}\right)}{\log 2}\\
f^{-1}(x) &= \frac{\log\left(\frac{1-3x}{3x-3}\right)}{\log 2}.
\end{aligned}$$

61. We start with $g(g^{-1}(x)) = x$ and substitute $y = g^{-1}(x)$. We have

$$\begin{aligned}
g(y) &= x\\
2^{\sin y} &= x\\
\ln\left(2^{\sin y}\right) &= \ln x\\
\ln(2^{\sin y}) = (\sin y)\ln 2 &= \ln x\\
\sin y &= \frac{\ln x}{\ln 2}\\
y &= \arcsin\left(\frac{\ln x}{\ln 2}\right).
\end{aligned}$$

Thus

$$g^{-1}(x) = \arcsin\left(\frac{\ln x}{\ln 2}\right).$$

65. False. Suppose $f(x) = x$ and $g(x) = 2x + 1$. Then $f(3) = 3$ and $g(1) = 3$, which means that $f(3) = g(1)$. But $g(3)$ does not equal $f(1)$, because $g(3) = 7$ and $f(1) = 1$.

69. False. Suppose $f(x) = x + 1$. Then $f(x^2) = x^2 + 1$, but $[f(x)]^2 = (x + 1)^2 = x^2 + 2x + 1$.

73. This is an increasing function, because as x increases, $g(x)$ decreases, and as $g(x)$ decreases, $g(g(x))$ increases.

77. (a) $f(8) = 2$, because 8 divided by 3 equals 2 with a remainder of 2. Similarly, $f(17) = 2$, $f(29) = 2$, and $f(99) = 0$.
   (b) $f(3x) = 0$ because, no matter what x is, $3x$ will be divisible by 3.
   (c) No. Knowing, for example, that $f(x) = 0$ tells us that x is evenly divisible by 3, but gives us no other information regarding x.
   (d) $f(f(x)) = f(x)$, because $f(x)$ equals either 0, 1, or 2, and $f(0) = 0$, $f(1) = 1$, and $f(2) = 2$.
   (e) No. For example, $f(1) + f(2) = 1 + 2 = 3$, but $f(1 + 2) = f(3) = 0$.

CHAPTER NINE

Solutions for Section 9.1

1. Larger powers of x give smaller values for $0 < x < 1$.
   A - (iii)
   B - (ii)
   C - (iv)
   D - (i)

5. (a) Note that $y = x^{-10} = (\frac{1}{x})^{10}$ is undefined at $x = 0$. Since $y = (\frac{1}{x})^{10}$ is raised to an even power, the graph "explodes" in the same direction as x approaches zero from the left and the right. Thus, as $x \longrightarrow 0, x^{-10} \longrightarrow +\infty, -x^{10} \longrightarrow 0$.
   (b) As $x \longrightarrow \infty$, $x^{-10} \longrightarrow 0$, $-x^{10} \longrightarrow -\infty$.
   (c) As $x \longrightarrow -\infty$, $x^{-10} \longrightarrow 0$, $-x^{10} \longrightarrow -\infty$.

9. (a) $x^{1/n}$ is concave down: its values increase quickly at first and then more slowly as x gets larger. The function x^n, on the other hand, is concave up. Its values increase at an increasing rate as x gets larger. Thus

$$f(x) = x^{1/n}$$

   and

$$g(x) = x^n.$$

   (b) Since the point A is the intersection of $f(x)$ and $g(x)$, we want the solution of the equation $x^n = x^{1/n}$. Raising both sides to the power of n we get

$$(x^n)^{\cdot n} = (x^{1/n})^{\cdot n}$$

   or in other words

$$x^{n^2} = x$$

   Since $x \neq 0$ at the point A, we can divide both sides by x, giving

$$x^{n^2-1} = 1.$$

   Since $n^2 - 1$ is just some integer we rewrite the equation as

$$x^p = 1$$

   where $p = n^2 - 1$. If p is even, $x = \pm 1$, if p is odd $x = 1$. By looking at the graph we can tell that we are not interested in the situation when $x = -1$. When $x = 1$, the quantities x^n and $x^{1/n}$ both equal 1. Thus the coordinates of point A are $(1, 1)$.

13. (a) We have

$$f(x) = g\left(h(x)\right) = 16x^4.$$

   Since $g(x) = 4x^2$, we know that

$$g\left(h(x)\right) = 4\left(h(x)\right)^2 = 16x^4$$

$$(h(x))^2 = 4x^4.$$

$$\text{Thus,} \quad h(x) = 2x^2 \text{ or } -2x^2.$$

   Since $h(x) \leq 0$ for all x, we know that

$$h(x) = -2x^2.$$

   (b) We have

$$f(x) = j\left(2g(x)\right) = 16x^4, \qquad j(x) \text{ a power function.}$$

   Since $g(x) = 4x^2$, we know that

$$j\left(2g(x)\right) = j(8x^2) = 16x^4.$$

Since $j(x)$ is a power function, $j(x) = kx^p$. Thus,

$$\begin{aligned} j(8x^2) = k(8x^2)^p &= 16x^4 \\ k \cdot 8^p x^{2p} &= 16x^4. \end{aligned}$$

Since $x^{2p} = x^4$ if $p = 2$, letting $p = 2$, we have

$$\begin{aligned} k \cdot 64x^4 &= 16 \cdot x^4 \\ 64k &= 16 \\ k &= \frac{1}{4} \end{aligned}$$

Thus, $j(x) = \frac{1}{4}x^2$.

17. Using

$$d = 483{,}000{,}000,$$

we have

$$P = 365\left(\frac{483{,}000{,}000}{93{,}000{,}000}\right)^{3/2}$$

which gives

$$P \approx 4320 \text{ earth days},$$

or almost 12 earth years.

Solutions for Section 9.2

1. (a) The function fits neither, because $h(x) = 3(-2)^{3x} = 3((-2)^3)^x = 3(-8)^x$, and the base of an exponential function must be positive.
   (b) The function can be written as an exponential, because $j(x) = 3(-3)^{2x} = 3((-3)^2)^x = 3 \cdot 9^x$.
   (c) The function fits neither form. If the expression in the parentheses expanded, then $m(x) = 3(9x^2 + 6x + 1) = 27x^2 + 18x + 3$.
   (d) The function is exponential, because $n(x) = 3 \cdot 2^{3x+1} = 3 \cdot 2^{3x} \cdot 2^1 = 6 \cdot 8^x$.
   (e) The function is exponential, because $p(x) = (5^x)^2 = 5^{2x} = (5^2)^x = 25^x$.
   (f) The function fits neither, because the variable in the exponent is squared.
   (g) The function fits an exponential, because $r(x) = 2 \cdot 3^{-2x} = 2(3^{-2})^x = 2(\frac{1}{9})^x$.
   (h) The function is a power function, because $s(x) = \frac{4}{5x^{-3}} = \frac{4}{5}x^3$.

5. (a) For $0 < x < 1$, we know that $x^3 < x^2$ and we know that for $x > 0$, $x^2 < 2x^2$. Therefore, $f(x)$ is graph (C), $g(x)$ is graph (A), $h(x)$ is graph (B).
   (b) Yes, (B) $= g(x) = h(x)$ and (A) $= x^3 = 2x^2$, so

$$\begin{aligned} x^3 &= 2x^2 \\ x^3 - 2x^2 &= 0 \\ x^2(x-2) &= 0 \end{aligned}$$

   So $x = 2$ is the only solution for $x > 0$.
   (c) No, since $2x^2 > x^2$ for $x > 0$.

9. We need to solve $j(x) = kx^p$ for p and k. We know that $j(x) = 2$ when $x = 1$. Since $j(1) = k \cdot 1^p = k$, we have $k = 2$. To solve for p, use the fact that $j(2) = 16$ and also $j(2) = 2 \cdot 2^p$, so

$$2 \cdot 2^p = 16,$$

giving $2^p = 8$, so $p = 3$. Thus, $j(x) = 2x^3$.

13. Since $f(1) = k \cdot 1^p = k$, we know $k = f(1) = \frac{3}{2}$
Since $f(2) = k \cdot 2^p = \frac{3}{8}$, and since $k = \frac{3}{2}$, we know

$$\left(\frac{3}{2}\right) \cdot 2^p = \frac{3}{8}$$

which implies

$$2^p = \frac{3}{8} \cdot \frac{2}{3} = \frac{1}{4}.$$

Thus $p = -2$, and $f(x) = \frac{3}{2} \cdot x^{-2}$.

17. (a) If f is linear,

$$m = \frac{128 - 16}{2 - 1} = 112,$$

and

$$16 = 112(1) + b \quad \Rightarrow \quad b = -96.$$

Thus,

$$f(x) = 112x - 96.$$

(b) If f is exponential, then

$$\frac{128}{16} = \frac{a(b)^2}{a(b)} = b \quad \Rightarrow \quad b = 8$$

and

$$16 = a(8) \quad \Rightarrow \quad a = 2.$$

Therefore

$$f(x) = 2(8)^x.$$

(c) If f is a power function, $f(x) = k(x)^p$. Then

$$\frac{f(2)}{f(1)} = \frac{k(2)^p}{k(1)^p} = (2)^p = \frac{128}{16} = 8,$$

so $p = 3$. Using $f(1) = 16$ to solve for k, we have

$$16 = k(1^3) \quad \Rightarrow \quad k = 16.$$

Thus,

$$f(x) = 16x^3.$$

21. $c(t) = \frac{1}{t}$ is indeed one possible formula. It is not, however, the only one. Because the vertical and horizontal axes are asymptotes for this function, we know that the power p is a negative number and

$$c(t) = kt^p.$$

If $p = -3$ then $c(t) = kt^{-3}$. Since $(2, \frac{1}{2})$ lies on the curve, $\frac{1}{2} = k(2)^{-3}$ or $k = 4$. So, $c(t) = 4t^{-3}$ could describe this function. Similarly, so could $c(t) = 16x^{-5}$ or $c(t) = 64x^{-7}$...

25. The function $f(d) = b \cdot d^{p/q}$, with $p < q$, because $f(d)$ increases more and more slowly as d gets larger, and $g(d) = a \cdot d^{p/q}$, with $p > q$, because $g(d)$ increases more and more quickly as d gets larger.

Solutions for Section 9.3

1. By multiplying out the expression $x(x - 3)(x + 2)$ and then simplifying the result, we see that

$$u(x) = x^3 - x^2 - 6x,$$

So u is a third-degree polynomial.

5. $y = 1 - 2x^4 + x^3$ is a fourth degree polynomial with three terms. Its long-run behavior is that of $y = -2x^4$: as $x \to \pm\infty$, $y \to -\infty$.

9. The graph of $y = g(x)$ is shown in Figure 9.1 on the window $-5 \le x \le 5$ by $-20 \le y \le 10$.

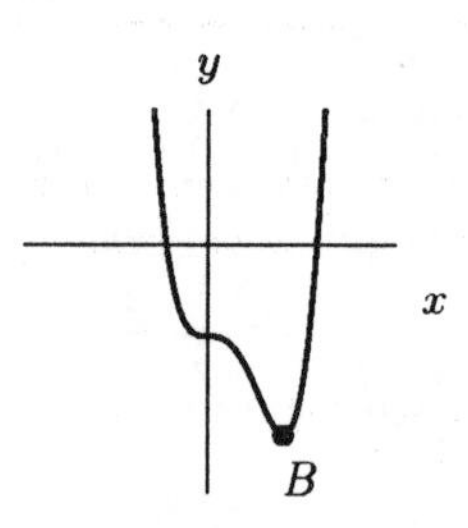

Figure 9.1

The minimum value of g occurs at point B as shown in the figure. Using either a table feature or trace on a graphing calculator, we approximate the minimum value of g to be -16.54 (to two decimal places).

13. Use a graphing calculator or computer to approximate values where $f(x) = g(x)$ or to find the zeros for $f(x) - g(x)$. In either case, we find the points of intersection for f and g to be $x \approx -1.764$, $x \approx 0.875$ and $x \approx 3.889$. The values of x for which $f(x) < g(x)$ are on the interval $-1.764 < x < 0.875$ or $x > 3.889$.

17. (a) We are interested in V for $0 \le T \le 30$, and the y-intercept of V occurs at (0, 999.87). If we look at the graph of V on the window $0 \le x \le 30$ by $0 \le y \le 1500$, the graph looks like a horizontal line. Since V is a cubic polynomial, we suspect more interesting behavior with a better choice of window. Note that $V(30) \approx 1003.77$, so we know V varies (at least) from $V = 999.87$ to $V \approx 1003.77$. Change the range to $998 \le y \le 1004$. On this window we see a more appropriate view of the behavior of V for $0 \le T \le 30$. [Note: To view the function V as a cubic, a much larger window is needed. Try $-500 \le x \le 500$ by $-3000 \le y \le 5000$.]

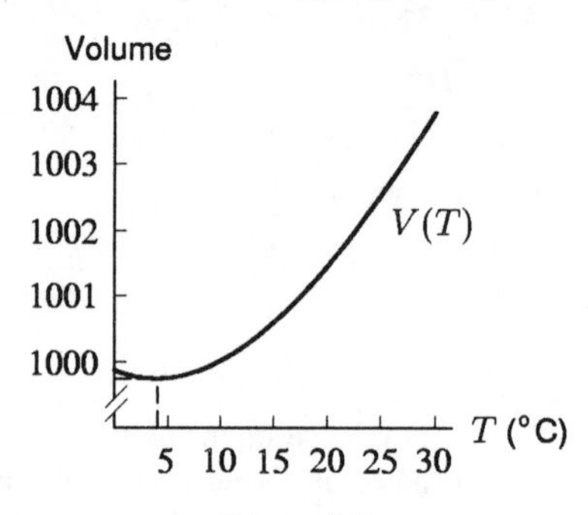

Figure 9.2

(b) The graph of V decreases for $0 \le T \le 3.96$ and then increases for $3.96 < T < 30$. The function is concave up on the interval $0 \le T \le 30$ (i.e., the graph bends upward). Thus, the volume of 1 kg of water decreases as T increases from 0° C to 3.96° C and increases thereafter. The volume increases at an increasing rate as the temperature increases.

(c) If density, d, is given by $d = m/V$ and m is constant, then the maximum density occurs when V is minimum. Thus, the maximum density occurs when $T \approx 3.96°C$. [Note: We have $m = 1$, but a graph of $y = 1/V$ is very difficult to distinguish from a horizontal line. One possible choice of window to view $y = 1/V$ is $0 \le x \le 30$, $0.000996 \le y \le 0.001001$.]

21. (a) False. For example, $f(x) = x^2 + x$ is not even.
(b) False. For example, $f(x) = x^2$ is not invertible.
(c) True.
(d) False. For example, if $f(x) = x^2$, then

$$f(x) \to \infty \quad \text{as} \quad x \to \infty$$
$$f(x) \to \infty \quad \text{as} \quad x \to -\infty.$$

Solutions for Section 9.4

1. The graph in the text represents a polynomial of even degree, degree at least 4. Zeros are shown at $x = -2$, $x = -1$, $x = 2$, and $x = 3$. The leading coefficient must be negative. Thus, of the choices in the table, only C and E are possibilities. When $x = 0$, function C gives

$$y = -\frac{1}{2}(2)(1)(-2)(-3) = -\frac{1}{2}(12) = -6,$$

and function E gives

$$y = -(2)(1)(-2)(-3) = -12.$$

Since the y-intercept appears to be $(0, -6)$ rather than $(0, -12)$, function C best fits the polynomial shown.

5. The graph of h shows zeros at $x = 0$, $x = 3$, and a repeated zero at $x = -2$. Thus

$$h(x) = x(x+2)^2(x-3).$$

Check by multiplying and gathering like terms.

9. We use the position of the "bounce" on the x-axis to indicate a repeated zero at that point. Since there is not a sign change at those points, the zero is repeated an even number of times. Thus, let

$$y = k(x+3)(x+1)(x-2)^2$$

for some $k > 0$. Since there is no scale on the y-axis and no coordinates are given for additional points on the graph, we do not have sufficient information to determine k.

13. We know that $g(-2) = 0$, $g(-1) = -3$, $g(2) = 0$, and $g(3) = 0$. We also know that $x = -2$ is a repeated zero. Thus, let

$$g(x) = k(x+2)^2(x-2)(x-3).$$

Then, using $g(-1) = -3$, gives

$$g(-1) = k(-1+2)^2(-1-2)(-1-3) = k(1)^2(-3)(-4) = 12k,$$

so $12k = -3$, and $k = -\frac{1}{4}$. Thus,

$$g(x) = -\frac{1}{4}(x+2)^2(x-2)(x-3)$$

is a possible formula for g.

17. This one is tricky. However, we can view j as a translation of another function. Consider the graph in Figure 9.3. A formula for the graph in Figure 9.3 could be of the form $y = k(x+3)(x+2)(x+1)$. Since $y = 6$ if $x = 0$, $6 = k(0+3)(0+2)(0+1)$, therefore $6 = 6k$, which yields $k = 1$. Note that the graph of $j(x)$ is a vertical shift (by 4) of the graph in Figure 9.3, giving $j(x) = (x+3)(x+2)(x+1) + 4$ as a possible formula for j.

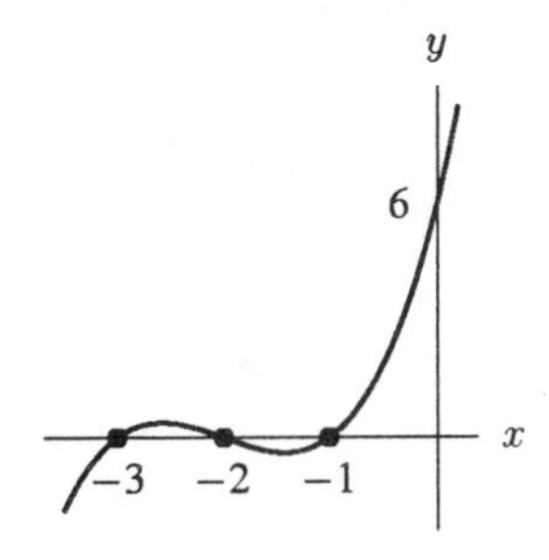

Figure 9.3

21. To pass through the given points, the polynomial must be of at least degree 2. Thus, let f be of the form

$$f(x) = ax^2 + bx + c.$$

Then using $f(0) = 0$ gives

$$a(0)^2 + b(0) + c = 0,$$

so $c = 0$. Then, with $f(2) = 0$, we have

$$\begin{aligned} a(2)^2 + b(2) + 0 &= 0 \\ 4a + 2b &= 0 \\ \text{so} \quad b &= -2a. \end{aligned}$$

Using $f(3) = 3$ and $b = -2a$ gives

$$a(3)^2 + (-2a)(3) + 0 = 3$$

so

$$\begin{aligned} 9a - 6a &= 3 \\ 3a &= 3 \\ a &= 1. \end{aligned}$$

Thus, $b = -2a$ gives $b = -2$. The unique polynomial of degree ≤ 2 which satisfies the given conditions is $f(x) = x^2 - 2x$.

25. The points $(-3,0)$ and $(1,0)$ indicate two zeros for the polynomial. Thus, the polynomial must be of at least degree 2. We could let $p(x) = k(x+3)(x-1)$ as in the previous problems, and then use the point $(0,-3)$ to solve for k. An alternative method would be to let $p(x)$ be of the form

$$p(x) = ax^2 + bx + c$$

and solve for a, b, and c using the given points.

The point $(0,-3)$ gives

$$\begin{aligned} a \cdot 0 + b \cdot 0 + c &= -3, \\ \text{so} \quad c &= -3. \end{aligned}$$

Using $(1,0)$, we have

$$\begin{aligned} a(1)^2 + b(1) - 3 &= 0 \\ \text{which gives} \quad a + b &= 3. \end{aligned}$$

The point $(-3,0)$ gives

$$\begin{aligned} a(-3)^2 + b(-3) - 3 &= 0 \\ 9a - 3b &= 3 \\ \text{or} \quad 3a - b &= 1. \end{aligned}$$

From $a + b = 3$, substitute

$$a = 3 - b$$

into

$$3a - b = 1.$$

Then

$$\begin{aligned} 3(3-b) - b &= 1 \\ 9 - 3b - b &= 1 \\ -4b &= -8 \\ \text{so} \quad b &= 2. \end{aligned}$$

Then $a = 3 - 2 = 1$. Therefore,

$$p(x) = x^2 + 2x - 3$$

is the polynomial of least degree through the given points.

29. $y = 4x^2 - 1 = (2x-1)(2x+1)$, which implies that $y = 0$ for $x = \pm\frac{1}{2}$.

33. (a) $V(x) = x(6-2x)(8-2x)$
(b) Values of x for which $V(x)$ makes sense are $0 < x < 3$, since if $x < 0$ or $x > 3$ the volume is negative.
(c) See Figure 9.4.
(d) Using a graphing calculator, we find the peak between $x = 0$ and $x = 3$ to occur at $x \approx 1.13$. The maximum volume is ≈ 24.26 in^3.

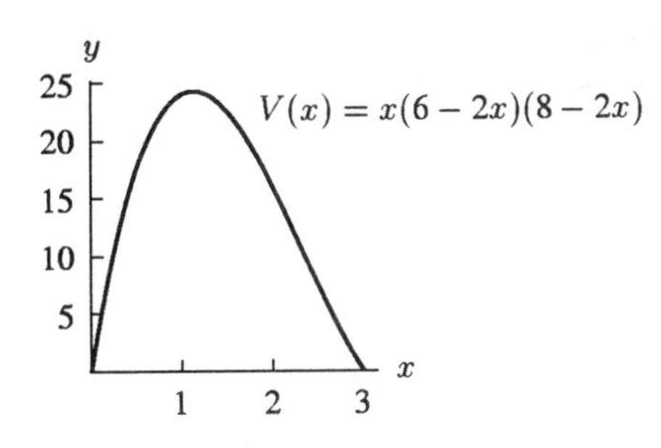

Figure 9.4

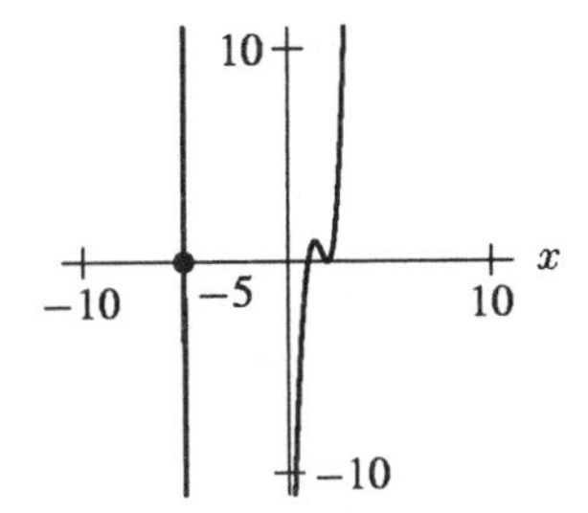

Figure 9.5: $f(x) = x^4 - 17x^2 + 36x - 20$

37. (a) On the standard viewing screen, the graph of f is shown in Figure 9.5:
(b) No. The graph of f is very steep near $x = -5$, but that doesn't mean it has a vertical asymptote. Since f is a polynomial function, it is defined for all values of x.
(c) The function has 3 zeros. A good screen to see the zeros is $-6 \le x \le 3, -3 \le y \le 3$.
(d) Using a graphing calculator or a computer, we find that f has zeros at $x = -5$, $x = 1$, and a double zero at $x = 2$. Thus, $f(x) = (x+5)(x-1)(x-2)^2$.
(e) The function has 3 turning points, two of which are visible in the standard viewing window (see Figure 9.5), and one of which is in the third quadrant, but off the bottom of the screen. It is not possible to see all the turning points in the same window. To see the left-most turning point, a good window is $-6 \le x \le 6, -210 \le y \le 50$, but the other turning points are invisible on this scale. To see the other turning points, a good window is $0 \le x \le 3$, $-1 \le y \le 2$, but the left-most turning point is far too low to see on this window.

Solutions for Section 9.5

1. (a)

x	2	2.5	2.75	2.9	2.99	3	3.01	3.1	3.25	3.5	4
$f(x)$	−1	−2	−4	−10	−100	undefined	100	10	4	2	1

As x approaches 3 from the left, $f(x)$ takes on very large negative values. As x approaches 3 from the right, $f(x)$ takes on very large positive values.

(b)

x	5	10	100	1000
$f(x)$	0.5	0.143	0.010	0.001

x	−5	−10	−100	−1000
$f(x)$	−0.125	−0.077	−0.010	−0.001

For $x > 3$, as x increases, $f(x)$ approaches 0 from above. For $x < 3$, as x decreases, $f(x)$ approaches 0 from below.

(c) The horizontal asymptote is $y = 0$ (the x-axis). The vertical asymptote is $x = 3$. See Figure 9.6.

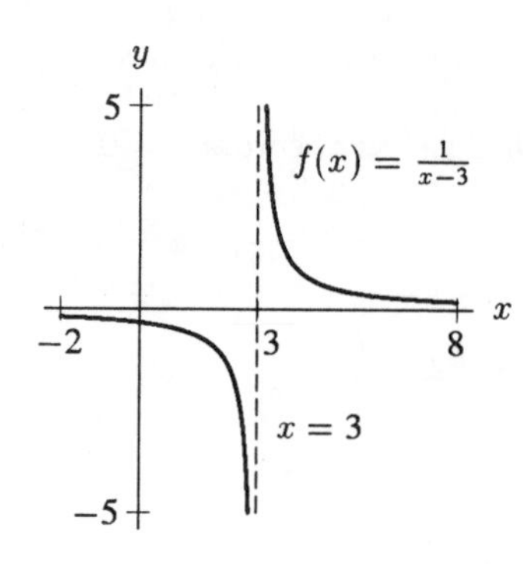

Figure 9.6

5. For the function f, $f(x) \to 1$ as $x \to \pm\infty$ since for large values of x, $f(x) \approx \frac{x^2}{x^2} = 1$.
   The function $g(x) \approx \frac{x^3}{x^2} = x$ for large values of x. Thus, as $x \to \pm\infty$, $g(x)$ approaches the line $y = x$.
   The function h will behave like $y = \frac{x}{x^2} = \frac{1}{x}$ for large values of x. Thus, $h(x) \to 0$ as $x \to \pm\infty$.

9. As $x \to \pm\infty$, $1/x \to 0$ and $x/(x+1) \to 1$, so $h(x)$ approaches $3 - 0 + 1 = 4$.
   Therefore $y = 4$ is the horizontal asymptote.

13. (a) Originally the total amount of the alloy is 2 kg, one half of which — or equivalently 1 kg — is tin. We have

$$\begin{aligned} C(x) &= \frac{\text{Total amount of tin}}{\text{Total amount of alloy}} \\ &= \frac{(\text{original amount of tin}) + (\text{added tin})}{(\text{original amount of alloy}) + (\text{added tin})} \\ &= \frac{1+x}{2+x} \end{aligned}$$

$C(x)$ is a rational function.

(b) Using our formula, we have

$$C(0.5) = \frac{1+0.5}{2+0.5} = \frac{1.5}{2.5} = 60\%.$$

This means that if 0.5 kg of tin is added, the concentration of tin in the resulting alloy will be 60%. As for $C(-0.5)$, we have

$$C(-0.5) = \frac{1-0.5}{2-0.5} = \frac{0.5}{1.5} \approx 33\%.$$

A negative x-value corresponds to the removal of tin from the original mixture, so the statement $C(-0.5) = 33\%$ would mean that removing 0.5 kg of tin results in an alloy that is 33% tin.

(c) To graph $y = C(x)$, let's see if we can represent the formula as a translation of a power function. We write

$$\begin{aligned} C(x) = \frac{x+1}{x+2} &= \frac{(x+2)-1}{x+2} \\ &= \frac{x+2}{x+2} - \frac{1}{x+2} \quad \text{(splitting the numerator)} \\ &= 1 - \frac{1}{x+2} \\ &= -\frac{1}{x+2} + 1. \end{aligned}$$

Thus the graph of C will resemble the graph of $f(x) = -\frac{1}{x}$ shifted two units to the left and then one unit up. Figure 9.7 shows this translation.

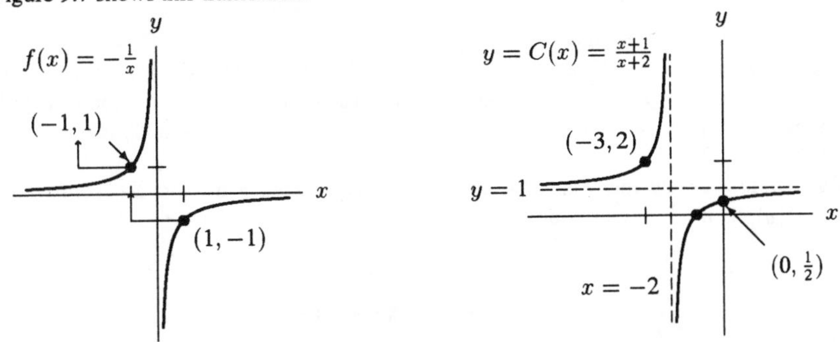

Figure 9.7: The graph of the rational function $y = C(x)$ is a translation of the graph of the power function $f(x) = -\frac{1}{x}$.

For "interesting features", we start with the intercepts and asymptotes. $C(x)$ has a y-intercept between $y = 0$ and $y = 1$, an x-intercept (or zero) at $x = -1$, a horizontal asymptote of $y = 1$, and a vertical asymptote of $x = -2$. What physical significance do these graphical features have?

First off, since

$$C(0) = \frac{0+1}{0+2} = \frac{1}{2} = 0.5$$

we see that the y-intercept is 0.5, or 50%. This means that if you add no tin (i.e. $x = 0$ kg), then the concentration is 50%, the original concentration of tin in the alloy.

Second, since

$$C(-1) = \frac{-1+1}{-1+2} = \frac{0}{1} = 0,$$

we see that the x-intercept is indeed at $x = -1$. This means that if you remove 1 kg of tin (i.e. $x = -1$ kg), then the concentration will be 0%, as there will be no tin remaining in the alloy.

This fact has a second implication: the graph of $C(x)$ is meaningless for $x < -1$, as it is impossible to remove more than 1 kg of tin. Thus, in the context of the problem at hand, the domain of $C(x)$ is $x \geq -1$. The graph on this domain is given by Figure 9.8. Notice that the vertical asymptote of the original graph (at $x = -2$) no longer appears, and it has no physical significance.

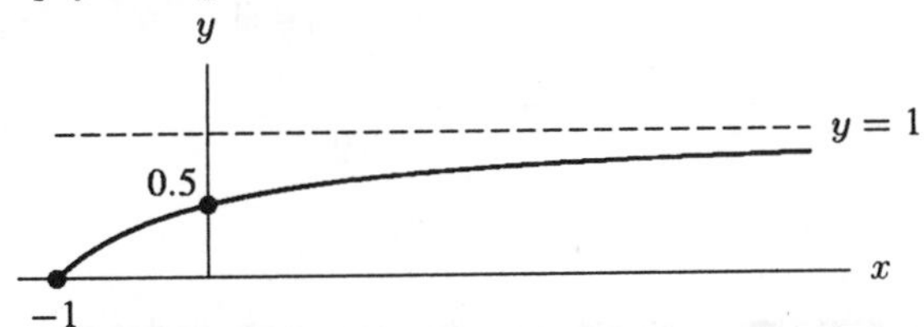

Figure 9.8: The domain of $C(x)$ is $x \geq -1$

The horizontal asymptote of $y = 1$ is, however, physically meaningful. As x grows large, we see that y approaches 1, or 100%. Since the amount of copper in the alloy is fixed at 1 kg, adding large amounts of tin results in an alloy that is nearly pure tin. For example, if we add 10 kg of tin then $x = 10$ and

$$C(x) = \frac{10+1}{10+2} = \frac{11}{12} = 0.916\ldots \approx 91.7\%.$$

Since the alloy now contains 11 kg of tin out of 12 kg total, it is relatively pure tin—at least, it is 91.7% pure. If instead we add 98 kg of tin, then $x = 98$ and

$$C(x) = \frac{98+1}{98+2} = \frac{99}{100} = 99\%.$$

Thus adding 98 kg of tin results in an alloy that is 99% pure. The 1 kg of copper is almost negligible. Therefore, the horizontal asymptote at $y = 1$ indicates that as the amount of added tin, x, grows large, the concentration of tin in the alloy approaches 1, or 100%.

17. (a) $C(x) = 30000 + 3x$
(b) $a(x) = \frac{C(x)}{x} = \frac{30000+3x}{x} = 3 + \frac{30000}{x}$
(c) The graph of $y = a(x)$ is shown in Figure 9.9.
(d) The average cost, $a(x)$, approaches \$3 per unit as the number of units grows large. This is because the fixed cost of \$30,000 is averaged over a very large number of goods, so that each good costs only little more than \$3 to produce.
(e) The average cost, $a(x)$, grows very large as $x \to 0$, because the fixed cost of \$30000 is being divided among a small number of units.
(f) The value of $a^{-1}(y)$ tells us how many units the firm must produce to reach an average cost of \$$y$ per unit. To find a formula for $a^{-1}(y)$, let $y = a(x)$, and solve for x. Then

$$\begin{aligned} y &= \frac{30000+3x}{x} \\ yx &= 30000 + 3x \\ yx - 3x &= 30000 \\ x(y-3) &= 30000 \\ x &= \frac{30000}{y-3}. \end{aligned}$$

So, we have $a^{-1}(y) = \frac{30000}{y-3}$.
(g) We want to evaluate $a^{-1}(5)$, the total number of units required to yield an average cost of \$5 per unit.

$$a^{-1}(5) = \frac{30000}{5-3} = \frac{30000}{2} = 15000.$$

Thus, the firm must produce at least 15,000 units for the average cost per unit to be \$5. The firm must produce at least 15,000 units to make a profit.

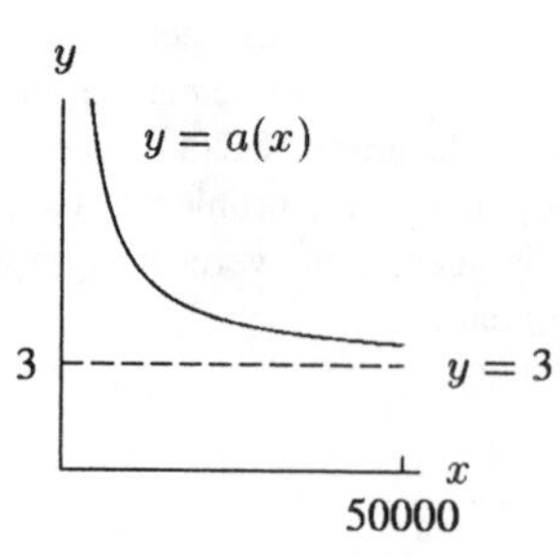

Figure 9.9

Solutions for Section 9.6

1. The zero of this function is at $x = -3$. It has a vertical asymptote at $x = -5$. Its long-run behavior is: $y \to 1$ as $x \to \pm\infty$.

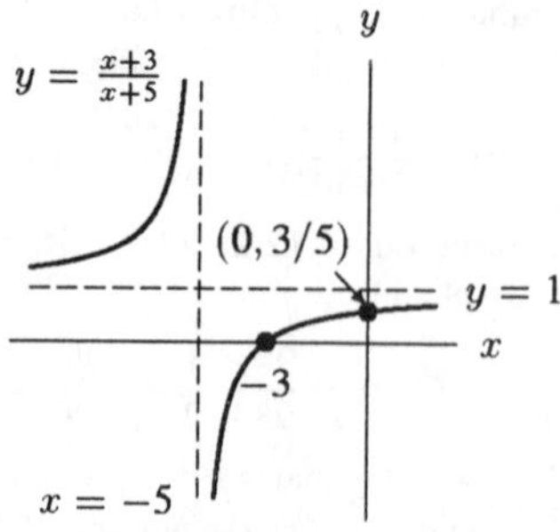

Figure 9.10

5. The graph will have vertical asymptotes at $x = \pm 4$ and zeros at $x = 3$ and $x = 2$. The y-intercept is $(0, -\frac{3}{4})$, and for large positive or negative values of x, we see that $y \to 2$—thus, there is a horizontal asymptote of $y = 2$. Note that the graph will intersect the horizontal asymptote if

$$\frac{2x^2 - 10x + 12}{x^2 - 16} = 2,$$

which implies

$$\begin{aligned} 2x^2 - 10x + 12 &= 2x^2 - 32 \\ -10x &= -44 \\ x &= 4.4 \end{aligned}$$

Putting all of this information together, we obtain a graph similar to that of Figure 9.11.

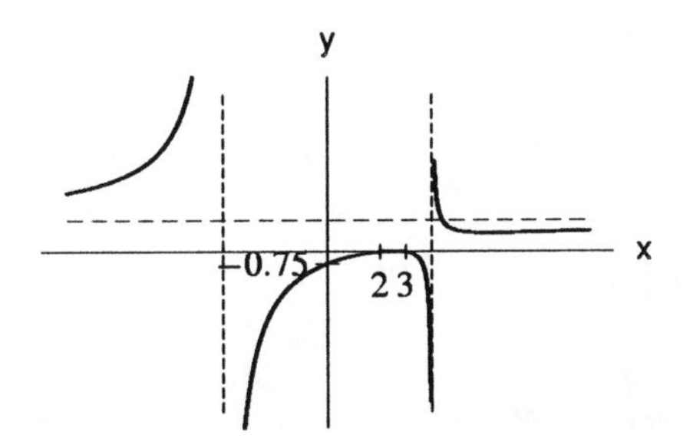

Figure 9.11

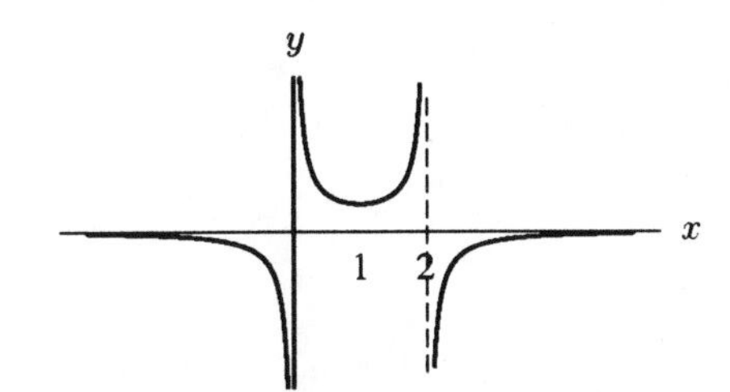

Figure 9.12

9. (a) The graph of $y = \frac{1}{f(x)}$ will have vertical asymptotes at $x = 0$ and $x = 2$. As $x \to 0$ from the left, $\frac{1}{f(x)} \to -\infty$, and as $x \to 0$ from the right, $\frac{1}{f(x)} \to +\infty$. The reciprocal of 1 is 1, so $\frac{1}{f(x)}$ will also go through the point (1,1). As $x \to 2$ from the left, $\frac{1}{f(x)} \to +\infty$, and as $x \to 2$ from the right, $\frac{1}{f(x)} \to -\infty$. As $x \to \pm\infty$, $\frac{1}{f(x)} \to 0$ and is negative.

   The graph of $y = \frac{1}{f(x)}$ is shown in Figure 9.12.

   (b) A formula for f is of the form

$$f(x) = k(x - 0)(x - 2) \qquad \text{and} \qquad f(1) = 1.$$

   Thus, $1 = k(1)(-1)$, so $k = -1$. Thus

$$f(x) = -x(x - 2).$$

   The reciprocal $\frac{1}{f(x)} = -\frac{1}{x(x-2)}$ is graphed as shown in Figure 9.12.

13. (a) The graph shows $y = 1/x$ shifted to the right one and up 2 units. Thus,

$$y = \frac{1}{x - 1} + 2$$

   is a choice for a formula.

   (b) The equation $y = 1/(x - 1) + 2$ can be written as

$$y = \frac{2x - 1}{x - 1}.$$

   (c) We see that the graph has both an x-and y-intercept. When $x = 0$, $y = \frac{-1}{-1} = 1$, so the y-intercept is $(0, 1)$. If $y = 0$ then $2x - 1 = 0$, so $x = \frac{1}{2}$. The x-intercept is $(\frac{1}{2}, 0)$.

17. The function f is the transformation of $y = \frac{1}{x}$, so $p = 1$. The graph of $y = \frac{1}{x}$ has been shifted three units to the right and four units up. To find the y-intercept, we need to evaluate $f(0)$:

$$f(0) = \frac{1}{-3} + 4 = \frac{11}{3}.$$

To find the x-intercepts, we need to solve $f(x) = 0$ for x.

$$\begin{aligned}
\text{Thus,} \quad 0 &= \frac{1}{x-3} + 4, \\
-4 &= \frac{1}{x-3}, \\
-4(x-3) &= 1, \\
-4x + 12 &= 1, \\
-4x &= -11, \\
\text{so} \quad x &= \frac{11}{4} \quad \text{is the only } x\text{-intercept.}
\end{aligned}$$

The graph of f is shown in Figure 9.13.

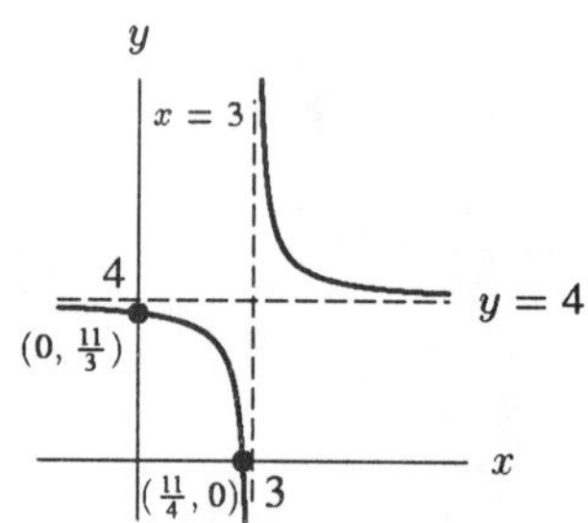

Figure 9.13

21. (a) The table indicates translation of $y = 1/x$ because the values of the function are headed in opposite directions near the vertical asymptote.

(b) The data points in the table indicate that $y \to \frac{1}{2}$ as $x \to \pm\infty$. The vertical asymptote does not appear to have been shifted. thus, we might try

$$y = \frac{1}{x} + \frac{1}{2}.$$

A check of x-values shows that this formula works. To express as a ratio of polynomials, we get a common denominator. Then

$$y = \frac{1(2)}{x(2)} + \frac{1(x)}{2(x)}$$

$$y = \frac{2+x}{2x}$$

25.
- Since the graph has an asymptote at $x = 2$, let the denominator be $(x - 2)$.
- Since the graph has a zero at $x = -1$, let the numerator be $(x + 1)$.
- Since the long–range behavior tends toward -1 as $x \to \pm\infty$, the ratio of the leading terms should be -1.

So a possible formula is $y = f(x) = -\left(\frac{x+1}{x-2}\right)$. You can check that the y–intercept is $y = \frac{1}{2}$, as it should be.

29. A guess of $y = \frac{(x-3)(x+2)}{(x+1)(x-2)}$ fits the zeros and vertical asymptote of the graph. However, in order to satisfy the y-intercept at $(0, -3)$ and end behavior of $y \to -1$ as $x \to \pm\infty$, the graph should be "flipped" across the x-axis. Thus try $y = -\frac{(x-3)(x+2)}{(x+1)(x-2)}$.

33. The vertical asymptotes indicate a denominator of $(x + 2)(x - 3)$. The horizontal asymptote of $y = 0$ indicates that the degree of the numerator is less than the degree of the denominator. To get the point (5,0) we need $(x - 5)$ as a factor in the numerator. Therefore, try

$$g(x) = \frac{(x-5)}{(x+2)(x-3)}.$$

37. Factoring the numerator, we have

$$g(x) = \frac{x^3 + 5x^2 + x + 5}{x + 5} = \frac{x^2(x+5) + (x+5)}{x+5} == \frac{(x+5)(x^2+1)}{(x+5)} = (x^2+1)\frac{(x+5)}{(x+5)}.$$

In this form we see that the graph of $y = g(x)$ is identical to that of $y = x^2 + 1$, except that the graph of $y = g(x)$ has no y-value corresponding to $x = -5$. The parabola $y = x^2 + 1$ goes through the point $(-5, 26)$ so the graph of $y = g(x)$ will be the parabola $y = x^2 + 1$ with a hole at $(-5, 26)$.

Solutions for Chapter 9 Review

1. Since the y-coordinates in the table fluctuate between 3 and -1, a trigonometric function might be a good model. Since at $x = 0$ the graph would be at a peak, a cosine seems appropriate. The amplitude is 2, the mid-line is $y = 1$ and the period is 2. Thus,

$$y = 2\cos(\pi x) + 1$$

is one possible choice. [Note: This answer is not the only possible choice.]

5. The data points in the table indicate an increasing non-linear function. In fact, the function values are increasing by greater and greater amounts as x increases. Try an exponential function. A look at the ratios of successive y-values shows a constant ratio of 5. The y-intercept of 0.5 indicates that

$$y = 0.5(5)^x$$

may be appropriate. In fact it fits beautifully!

9. Let $g(x) = k(x+2)(x^2)(x-2)$, since g has zeros at $x = \pm 2$, and a double zero at $x = 0$. Since $g(1) = 1$, we have $k(1+2)(1^2)(1-2) = 1$; thus $-3k = 1$ and $k = -\frac{1}{3}$. So

$$g(x) = -\frac{1}{3}(x^2)(x+2)(x-2)$$

is a possible formula.

13. To obtain the flattened effect of the graph near $x = 0$, let $x = 0$ be a multiple zero (of odd multiplicity). Thus, a possible choice would be $f(x) = kx^3(x+1)(x-2)$ for $k > 0$.

17. Suppose the earth's radius is R. We know that at the surface of the earth, $w = 150$, so

$$150 = \frac{k}{R^2}.$$

The person's weight is a gravitational force. If the earth's radius were to shrink, the person's weight would increase. We want to find the radius, r, that makes the weight equal to 1 ton, that is $w = 2000$, since 1 ton = 2000 lbs. Thus, we want the value of r satisfying:

$$2000 = \frac{k}{r^2}.$$

Dividing these two equations gives:

$$\frac{150}{2000} = \frac{k/R^2}{k/r^2} = \frac{r^2}{R^2}.$$

Solving for r gives:

$$r^2 = \frac{150}{2000}R^2$$

$$r = \sqrt{\frac{150}{2000}} \cdot R = 0.274R = 27.4\%R.$$

Thus, if the radius of the earth were to shrink to 27.4% of its current value, the person would not survive the gravitational force of his/her own weight.

21. (a) If $y = f(x)$, then $x = f^{-1}(y)$. Solving $y = f(x)$ for x, we have

$$\begin{aligned} y &= \frac{x}{x+5} \\ y(x+5) &= x \\ yx + 5y &= x \\ yx - x &= -5y \\ x(y-1) &= -5y \\ x &= \frac{-5y}{y-1} = \frac{5y}{1-y}. \end{aligned}$$

Thus, $f^{-1}(x) = 5x/(1-x)$.

(b) $f^{-1}(0.2) = \frac{5(0.2)}{(1-0.2)} = \frac{1}{0.8} = 1.25$. This means that 1.25 gallons of alcohol must be added to give an alcohol concentration of .20 or 20%.

(c) $f^{-1}(x) = 0$ means that $\dfrac{5x}{1-x} = 0$ which means that $x = 0$. This means that 0 gallons of alcohol must be added to give a concentration of 0%.

(d) The horizontal asymptote of $f^{-1}(x)$ is $y = -5$. Since x is a concentration of alcohol, $0 \le x \le 1$. Thus, the regions of the graph for which $x < 0$ and $x > 1$ have no physical significance. Consequently, since $f^{-1}(x)$ approaches its asymptote only as $x \to \pm\infty$, its horizontal asymptote has no physical significance.

25. $A < 0 < C < D < B$

29.
$$f(x) = \frac{p(x)}{q(x)} = \frac{-3(x-2)(x-3)}{(x-5)^2}$$

We need the factor of -3 in the numerator and the exponent of 2 in the denominator, because we have a horizontal asymptote of $y = -3$. The ratio of highest term of $p(x)$ to highest term of $q(x)$ will be $\frac{-3x^2}{x^2} = -3$.

33. (a) $-\frac{3}{r}$ is negative for $r > 0$; therefore, it is an attractive term. Similarly, $\frac{1}{r^2}$ is positive for $r > 0$; therefore, it is a repulsive term.

(b)

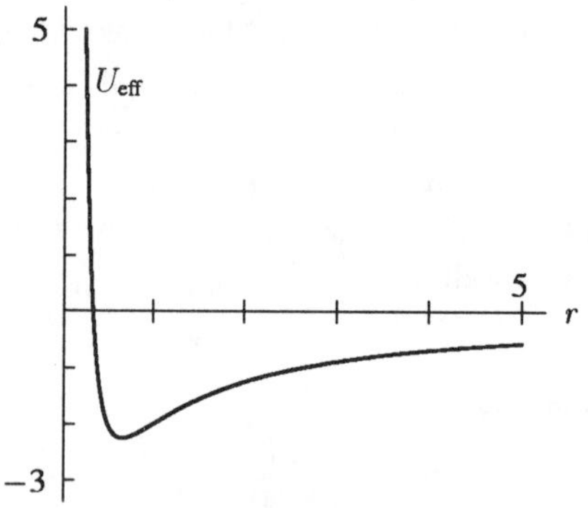

Figure 9.14: The effective potential, U_{eff}.

(c) The planet is most strongly attracted to the sun where U_{eff} is most negative. Using a computer or a graphing calculator, this seems to be where $r = 0.67$, or $r = \frac{2}{3}$.

(d) The horizontal asymptote of $y = U_{\text{eff}}$ is $y = 0$. Thus, as $r \to \infty$, the value of U_{eff} is negative but very close to zero. This means that a planet far from the sun will feel a very slight attraction towards the sun (due to gravity).

(e) If $y = U_{\text{eff}}$, then as $r \to 0, y \to \infty$ and the vertical asymptote is $r = 0$. This means that a planet in motion about the sun will experience a very strong repulsive "centrifugal force" as it draws near the sun. This is similar to the repelling force one experiences on a moving carousel when one tries to approach the center.

37. (a) With $u = 9$ and $l = 225$ we find $k = 9/\sqrt{225} = 3/5$. With $k = 0.6$ and $l = 4$, we find $u = (0.6)\sqrt{4} = 1.2$ meters/sec.

(b) Suppose the existing ship has speed u and length l, so

$$u = k\sqrt{l}.$$

The new ship has speed increased by 10%, so the new speed is $1.1u$. If the new length is L, since the constant remains the same, we have

$$1.1u = k\sqrt{L}.$$

Dividing these two equations we get

$$\frac{1.1u}{u} = \frac{k\sqrt{L}}{k\sqrt{l}}.$$

Simplifying and squaring we get

$$1.1 = \frac{\sqrt{L}}{\sqrt{l}}$$

$$(1.1)^2 = \frac{L}{l}$$

so

$$L = (1.1)^2 l = 1.21l.$$

Thus, the new hull length should be 21% longer than the hull length of the existing ship.

CHAPTER TEN

Solutions for Section 10.1

1. Scalar

5.
$$\vec{p} = 2\vec{w}, \quad \vec{q} = -\vec{u}, \quad \vec{r} = \vec{w} + \vec{u} = \vec{u} + \vec{w},$$
$$\vec{s} = \vec{p} + \vec{q} = 2\vec{w} - \vec{u}, \quad \vec{t} = \vec{u} - \vec{w}$$

9.

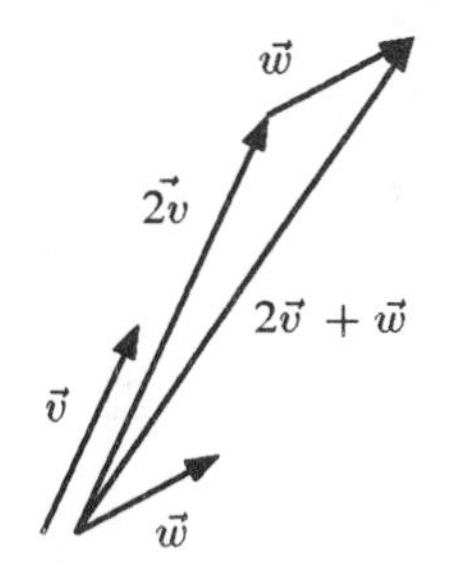

Figure 10.1

13.

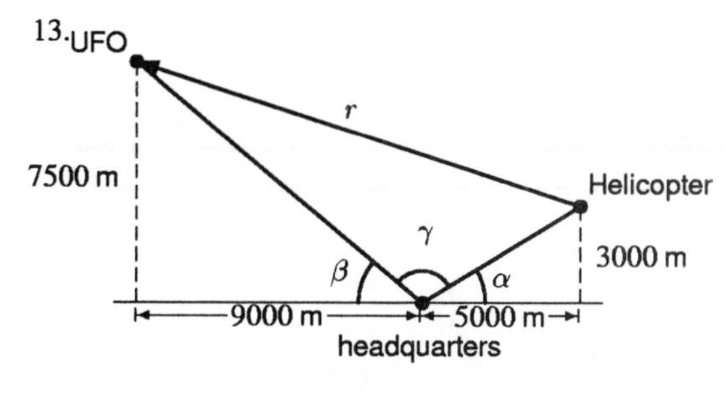

Figure 10.2

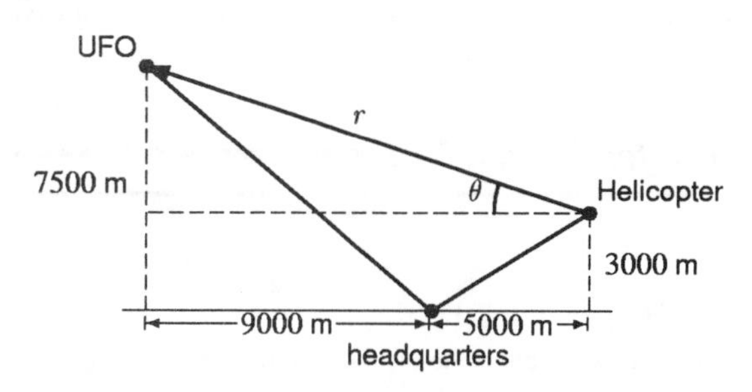

Figure 10.3

Figure 10.2 shows the headquarters at the origin, and a positive y-value as up, and a positive x-value as east. To solve for r, we must first find γ:

$$\begin{aligned}\gamma &= 180^\circ - \alpha - \beta \\ &= 180^\circ - \arctan\frac{3000}{5000} - \arctan\frac{7500}{9000} \\ &= 109.23^\circ.\end{aligned}$$

We now can find r using the Law of Cosines in the triangle formed by the position of the headquarters, the helicopter and the UFO.

In kilometers:

$$\begin{aligned}r^2 &= 34 + 137.25 - 2\cdot\sqrt{34}\cdot\sqrt{137.25}\cdot\cos\gamma \\ r^2 &= 216.32 \\ r &= 14.71\text{ km} \\ &= 14{,}710\text{ m}.\end{aligned}$$

From Figure 10.3 we see:

$$\begin{aligned}\tan\theta &= \frac{4500}{14{,}000} \\ \theta &= 17.82^\circ.\end{aligned}$$

Therefore, the helicopter must fly 14,710 meters with an angle of 17.82° from the horizontal.

17. The effect of scaling the left-hand picture in Figure 10.4 is to stretch each vector by a factor of a (shown with $a > 1$). Since, after scaling up, the three vectors $a\vec{v}$, $a\vec{w}$, and $a(\vec{v}+\vec{w})$ form a similar triangle, we know that $a(\vec{v}+\vec{w})$ is the sum of the other two: that is

$$a(\vec{v}+\vec{w}) = a\vec{v} + a\vec{w}.$$

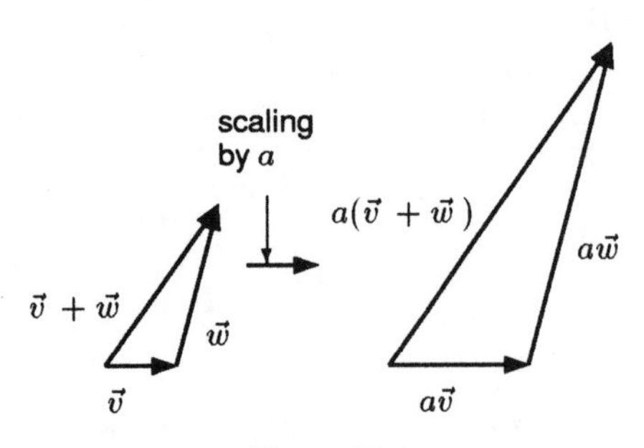

Figure 10.4

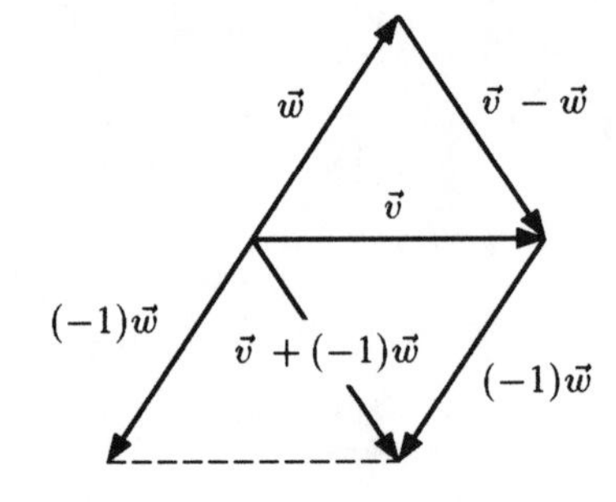

Figure 10.5

21. By Figure 10.5, the vectors $\vec{v}+(-1)\vec{w}$ and $\vec{v}-\vec{w}$ are equal.

Solutions for Section 10.2

1. $4\vec{i}+2\vec{j}-3\vec{i}+\vec{j}=\vec{i}+3\vec{j}$

5. The vector we want is the displacement from Q to P, which is given by

$$\overrightarrow{QP} = (1-4)\vec{i} + (2-6)\vec{j} = -3\vec{i} - 4\vec{j}.$$

9.

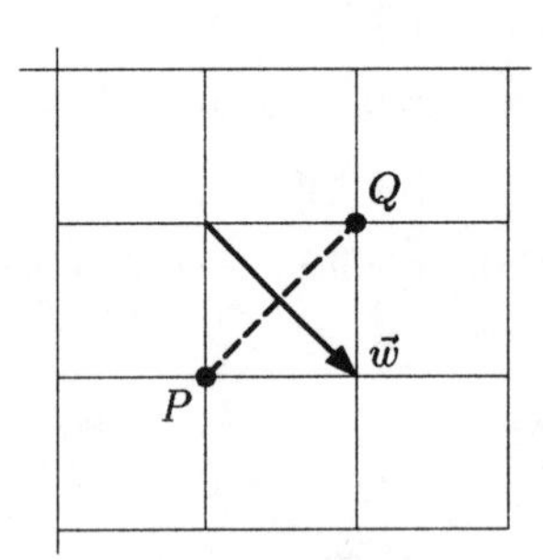

Figure 10.6

Figure 10.6 shows the vector $\vec{w}$ redrawn to show that it is perpendicular to the displacement vector $\overrightarrow{PQ}$, which lies along the dotted line. Thus, the angle is 90° or $\pi/2$.

13. $\|\vec{v}\| = \sqrt{7.2^2+(-1.5)^2+2.1^2} = \sqrt{58.5} \approx 7.6$

17. (a) If the car is going east, it is going solely in the positive x direction, so its velocity vector is $50\vec{i}$.
(b) If the car is going south, it is going solely in the negative y direction, so its velocity vector is $-50\vec{j}$.
(c) If the car is going southeast, the angle between the x-axis and the velocity vector is $-45°$. Therefore

$$\begin{aligned}\text{velocity vector} &= 50\cos(-45°)\vec{i} + 50\sin(-45°)\vec{j}\\ &= 25\sqrt{2}\vec{i} - 25\sqrt{2}\vec{j}.\end{aligned}$$

(d) If the car is going northwest, the velocity vector is at a 45° angle to the y-axis, which is 135° from the x-axis. Therefore:

$$\text{velocity vector} = 50(\cos 135°)\vec{i} + 50(\sin 135°)\vec{j} = -25\sqrt{2}\vec{i} + 25\sqrt{2}\vec{j}.$$

21. We get displacement by subtracting the coordinates of the origin $(0, 0, 0)$ from the coordinates of the cat $(1, 4, 0)$, giving Displacement$= (1-0)\vec{i} + (4-0)\vec{j} + (0-0)\vec{k} = \vec{i} + 4\vec{j}$.

Solutions for Section 10.3

1. (a) See the sketch in Figure 10.7, where $\vec{v}$ represents the first part of the man's walk, and $\vec{w}$ represents the second part.

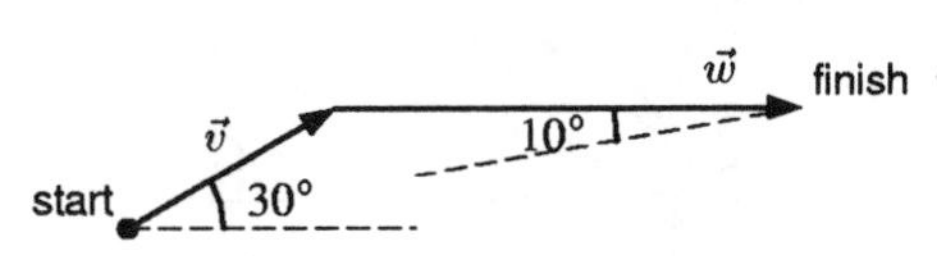

Figure 10.7

Since the man first walks 5 miles, we know $\|\vec{v}\| = 5$. Since he walks 30° north of east, resolving gives

$$\vec{v} = 5\cos 30°\vec{i} + 5\sin 30°\vec{j} = 4.33\vec{i} + 2.5\vec{j}.$$

For the second leg of his journey, the man walks a distance x miles due east, so $\vec{w} = x\vec{i}$.

(b) The vector from finish to start is $-(\vec{v} + \vec{w}) = -(4.33 + x)\vec{i} - 2.5\vec{j}$. This vector is at an angle of 10° south of west. So, using the magnitudes of the sides in the triangle in Figure 10.8:

$$\begin{aligned}\frac{2.5}{4.33 + x} &= \tan(10°) = 0.176\\ 2.5 &= 0.176(4.33 + x)\\ x &= \frac{2.5 - 0.176\cdot 4.33}{0.176} = 9.87.\end{aligned}$$

This means that $x = 9.87$.

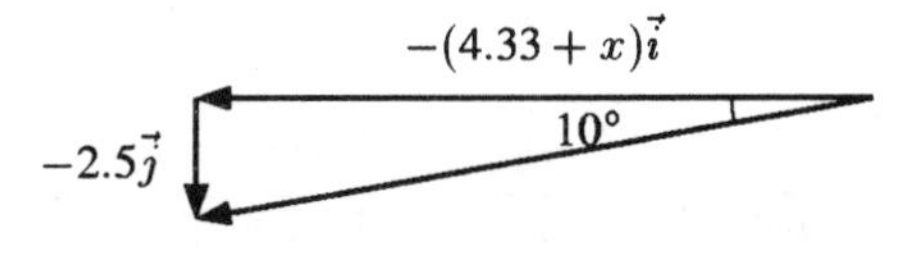

Figure 10.8

(c) The distance from the starting point is $\| -(4.33 + 9.87)\vec{i} - (2.5)\vec{j}\| = \sqrt{14.20^2 + 2.5^2} = 14.42$ miles.

5. Suppose $\vec{u}$ represents the velocity of the plane relative to the air and $\vec{w}$ represents the velocity of the wind. We can add these two vectors by adding their components. Suppose north is in the y-direction and east is the x-direction. The vector representing the airplane's velocity makes an angle of 45° with north; the components of $\vec{u}$ are

$$\vec{u} = 700\sin 45°\vec{i} + 700\cos 45°\vec{j} \approx 495\vec{i} + 495\vec{j}.$$

Since the wind is blowing from the west, $\vec{w} = 60\vec{i}$. By adding these we get a resultant vector $\vec{v} = 555\vec{i} + 495\vec{j}$. The direction relative to the north is the angle θ shown in Figure 10.9 given by

$$\begin{aligned}\theta &= \tan^{-1}\frac{x}{y} = \tan^{-1}\frac{555}{495}\\ &\approx 48.3°.\end{aligned}$$

The magnitude of the velocity is

$$\begin{aligned}\|\vec{v}\| &= \sqrt{495^2 + 555^2} = \sqrt{553{,}050}\\ &= 744 \text{ km/hr}.\end{aligned}$$

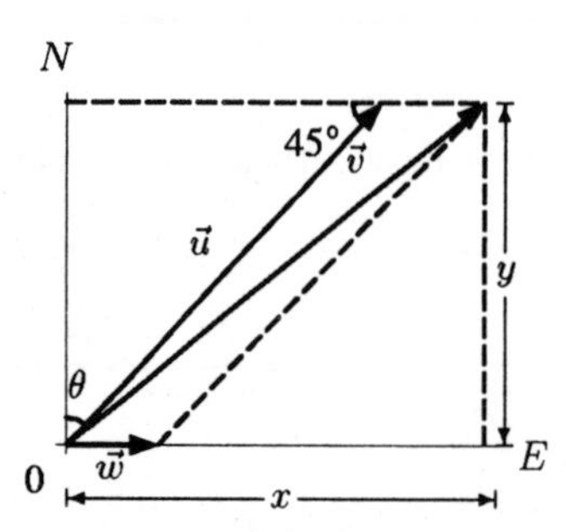

Figure 10.9: Note that θ is the angle between north and the vector $\vec{v}$

Solutions for Section 10.4

1. $\vec{c} \cdot \vec{y} = (\vec{i} + 6\vec{j}) \cdot (4\vec{i} - 7\vec{j}) = (1)(4) + (6)(-7) = 4 - 42 = -38$

5. Since $\vec{a} \cdot \vec{y}$ and $\vec{c} \cdot \vec{z}$ are both scalars, the answer to this equation is the product of two numbers and therefore a number. We have

$$\begin{aligned}\vec{a} \cdot \vec{y} &= (2\vec{j} + \vec{k}) \cdot (4\vec{i} - 7\vec{j}) = 0(4) + 2(-7) + 1(0) = -14\\ \vec{c} \cdot \vec{z} &= (\vec{i} + 6\vec{j}) \cdot (\vec{i} - 3\vec{j} - \vec{k}) = 1(1) + 6(-3) + 0(-1) = -17.\end{aligned}$$

Thus,

$$(\vec{a} \cdot \vec{y})(\vec{c} \cdot \vec{z}) = 238.$$

9.

$$\begin{aligned}\cos\theta &= \frac{(\vec{i} + \vec{j} + \vec{k}) \cdot (\vec{i} - \vec{j} - \vec{k})}{\|\vec{i} + \vec{j} + \vec{k}\|\|\vec{i} - \vec{j} - \vec{k}\|} = \frac{(1)(1) + (1)(-1) + (1)(-1)}{\sqrt{1^1 + 1^2 + 1^2}\sqrt{1^2 + (-1)^2 + (-1)^2}}\\ &= -\frac{1}{3}\end{aligned}$$

So, $\theta = \arccos(-\frac{1}{3}) \approx 1.91$ radians, or $\approx 109.5°$.

13.

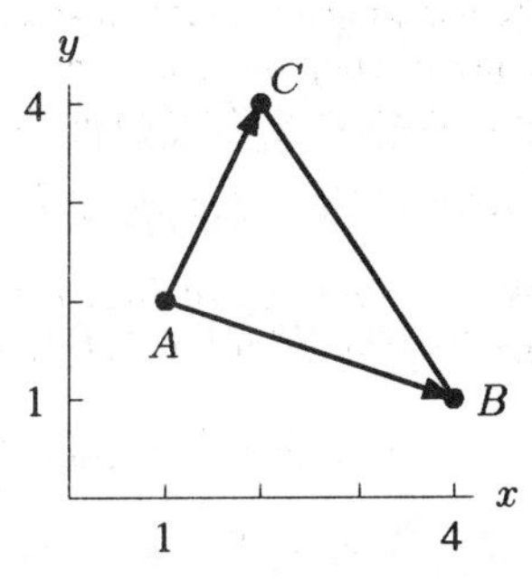

Figure 10.10

It is clear from the Figure 10.10 that only angle $\angle CAB$ could possibly be a right angle. Subtraction of x, y values for the points gives $\overrightarrow{AB} = 3\vec{i} - \vec{j}$ and $\overrightarrow{AC} = 1\vec{i} + 2\vec{j}$. Taking the dot product yields $\overrightarrow{AB} \cdot \overrightarrow{AC} = (3)(1) + (-1)(2) = 1$. Since this is non-zero, the angle can not be a right angle.

17. We have

$$\|\vec{a}_2\| = \sqrt{0.10^2 + 0.08^2 + 0.12^2 + 0.69^2} = 0.7120$$
$$\|\vec{a}_3\| = \sqrt{0.20^2 + 0.06^2 + 0.06^2 + 0.66^2} = 0.6948$$
$$\|\vec{a}_4\| = \sqrt{0.22^2 + 0.00^2 + 0.20^2 + 0.57^2} = 0.6429$$

$$\vec{a}_2 \cdot \vec{a}_3 = 0.10 \cdot 0.20 + 0.08 \cdot 0.06 + 0.12 \cdot 0.06 + 0.69 \cdot 0.66 = 0.4874$$
$$\vec{a}_3 \cdot \vec{a}_4 = 0.20 \cdot 0.22 + 0.06 \cdot 0.00 + 0.06 \cdot 0.20 + 0.66 \cdot 0.57 = 0.4322.$$

The distance between the English and the Bantus is given by θ where

$$\cos\theta = \frac{\vec{a}_2 \cdot \vec{a}_3}{\|\vec{a}_2\|\|\vec{a}_3\|} = \frac{0.4874}{(0.7120)(0.6948)} \approx 0.9852$$

so $\theta \approx 9.9°$.
The distance between the English and the Koreans is given by ϕ where

$$\cos\phi = \frac{\vec{a}_3 \cdot \vec{a}_4}{\|\vec{a}_3\|\|\vec{a}_4\|} = \frac{0.4322}{(0.6948)(0.6429)} \approx 0.9676$$

so $\phi \approx 14.6°$. Hence the English are genetically closer to the Bantus than to the Koreans.

Solutions for Chapter 10 Review

1. $-4\vec{i} + 8\vec{j} - 0.5\vec{i} + 0.5\vec{k} = -4.5\vec{i} + 8\vec{j} + 0.5\vec{k}$
5. $\|\vec{u}\| = \sqrt{1^2 + 1^2 + 2^2} = \sqrt{6}, \|\vec{v}\| = \sqrt{(-1)^2 + 2^2} = \sqrt{5}$
9. The velocity vector of the plane with respect to the calm air has the form

$$\vec{v} = a\vec{i} + 80\vec{k} \text{ where } \|\vec{v}\| = 480.$$

(See Figure 10.11.) Therefore $\sqrt{a^2 + 80^2} = 480$ so $a = \sqrt{480^2 - 80^2} \approx 473.3$ km/hr. We conclude that $\vec{v} \approx 473.3\vec{i} + 80\vec{k}$.

The wind vector is

$$\vec{w} = 100(\cos 45°)\vec{i} + 100(\sin 45°)\vec{j}$$
$$\approx 70.7\vec{i} + 70.7\vec{j}.$$

The velocity vector of the plane with respect to the ground is then

$$\begin{aligned}\vec{v} + \vec{w} &= (473.3\vec{i} + 80\vec{k}) + (70.7\vec{i} + 70.7\vec{j}) \\ &= 544\vec{i} + 70.7\vec{j} + 80\vec{k}.\end{aligned}$$

From Figure 10.12, we see that the velocity relative to the ground is

$$544\vec{i} + 70.7\vec{j}.$$

The ground speed is therefore $\sqrt{544^2 + 70.7^2} \approx 548.6$ km/hr.

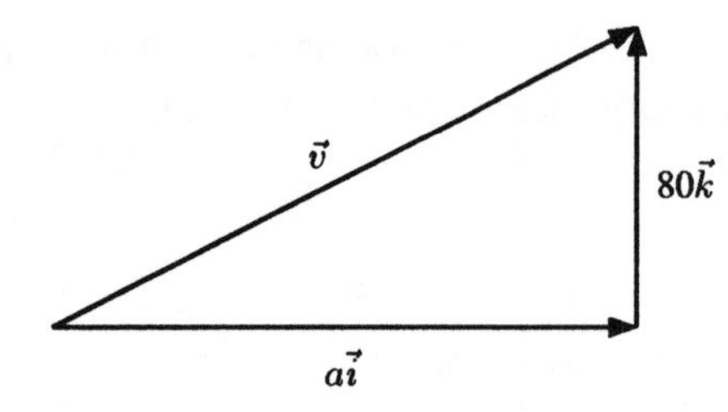

Figure 10.11: Side view

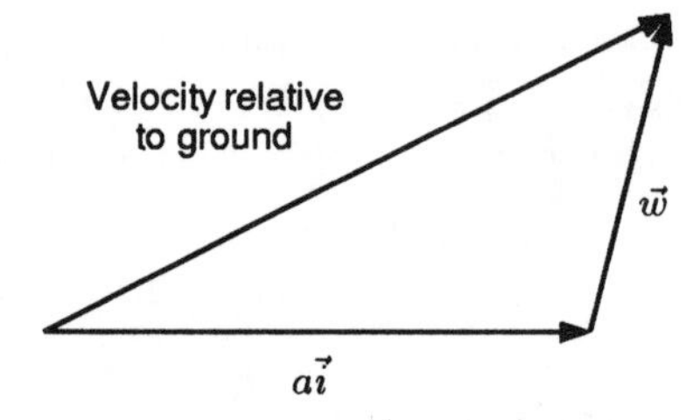

Figure 10.12: Top view

13.

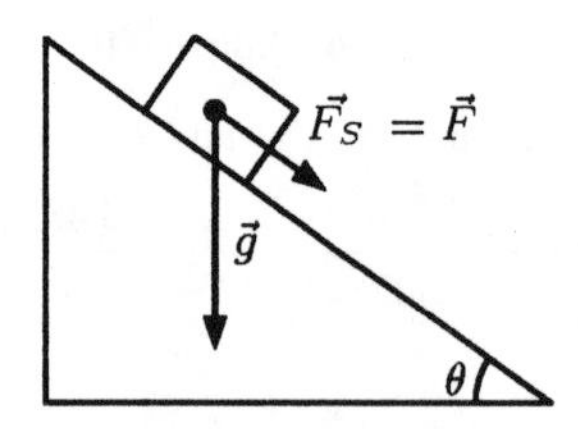

Figure 10.13

If $\theta = 0$ (the plank is at ground level), the sliding force is $F = 0$.
If $\theta = \pi/2$ (the plank is vertical), the sliding force equals g, the force due to gravity.
Therefore, we can guess that F is proportional with $\sin\theta$:

$$F = g\sin\theta.$$

This agrees with the bounds at $\theta = 0$ and $\theta = \pi/2$, and with the fact that the sliding force is smaller than g between 0 and $\pi/2$.

17. Suppose θ is the angle between $\vec{u}$ and $\vec{v}$.

(a) By the definition of scalar multiplication, we know that $-\vec{v}$ is in the opposite direction of $\vec{v}$, so the angle between $\vec{u}$ and $-\vec{v}$ is $\pi - \theta$. (See Figure 10.14.) Hence,

$$\begin{aligned}\vec{u} \cdot (-\vec{v}) &= \|\vec{u}\|\|-\vec{v}\|\cos(\pi - \theta) \\ &= \|\vec{u}\|\|\vec{v}\|(-\cos\theta) \\ &= -(\vec{u} \cdot \vec{v}).\end{aligned}$$

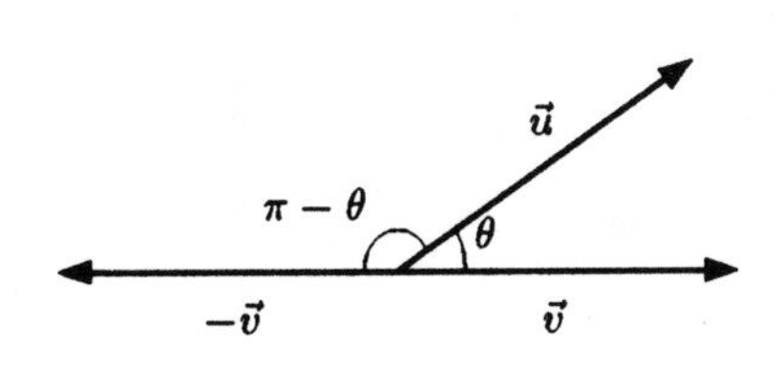

Figure 10.14

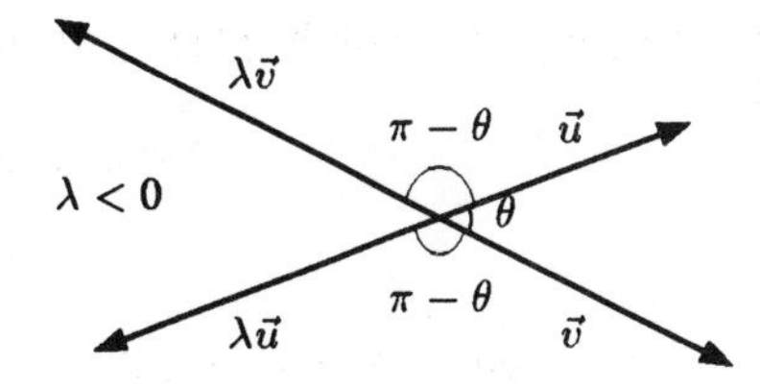

Figure 10.15

(b) If $\lambda < 0$, the angle between $\vec{u}$ and $\lambda\vec{v}$ is $\pi - \theta$, and so is the angle between $\lambda\vec{u}$ and $\vec{v}$. (See Figure 10.15.) So we have,

$$\begin{aligned}\vec{u} \cdot (\lambda\vec{v}) &= \|\vec{u}\|\|\lambda\vec{v}\|\cos(\pi - \theta)\\ &= |\lambda|\|\vec{u}\|\|\vec{v}\|(-\cos\theta)\\ &= -\lambda\|\vec{u}\|\|\vec{v}\|(-\cos\theta) \quad \text{since } |\lambda| = -\lambda\\ &= \lambda\|\vec{u}\|\|\vec{v}\|\cos\theta\\ &= \lambda(\vec{u} \cdot \vec{v}).\end{aligned}$$

By a similar argument, we have

$$\begin{aligned}(\lambda\vec{u}) \cdot \vec{v} &= \|\lambda\vec{u}\|\|\vec{v}\|\cos(\pi - \theta)\\ &= -\lambda\|\vec{u}\|\|\vec{v}\|(-\cos\theta)\\ &= \lambda(\vec{u} \cdot \vec{v}).\end{aligned}$$

21. Using the result of Problem 20, we have $\overrightarrow{AC} = \vec{w} + \vec{n} - \vec{m} = 3\vec{n} - 3\vec{m}$; $\overrightarrow{AB} = \vec{v} + \vec{m} + \vec{n} = 3\vec{m} + \vec{n}$; $\overrightarrow{AD} = \vec{v} + \vec{m} - (\vec{n} - \vec{m}) = 4\vec{m} - \vec{n}$; $\overrightarrow{BD} = (-\vec{n}) - (\vec{n} - \vec{m}) = \vec{m} - 2\vec{n}$

CHAPTER ELEVEN

Solutions for Section 11.1

1. The first term is $3(1)$, the second term is $3(2)$, up to $3(7)$. Thus, possible answer is
$$\sum_{n=1}^{7} 3n.$$

5. We have $n = 9$, $a_1 = 7$, $d = 7$. Using our formula, $S_n = (1/2)n(2a_1 + (n-1)d)$, we have
$$S_9 = \frac{9}{2}(2(7) + (9-1)7) = 315.$$

9. $\sum_{n=0}^{15}(2 + \frac{1}{2}n) = 2 + \frac{5}{2} + 3 + \frac{7}{2} + 4 + \cdots$. This is an arithmetic series with 16 terms (we start at $n = 0$ and continue to $n = 15$). Thus,
$$S_{16} = \frac{1}{2}(16)(2(2) + 15(1/2)) = 92.$$

13. We have $a_1 = 16$ and $d = 32$. The distance fallen by the object in n seconds is the sum
$$S_n = \frac{1}{2}n\left(2 \cdot 16 + (n-1)32\right) = \frac{1}{2}n(32 + 32n - 32) = \frac{1}{2}n(32n) = 16n^2.$$
So
$$f(n) = 16n^2.$$

17. (a) See Table 11.1.

TABLE 11.1 *Marginal cost for producing furniture*

n	c	Δc	n	c	Δc
1	300	−20	7	225	−2
2	280	−17	8	223	1
3	263	−14	9	224	4
4	249	−11	10	228	7
5	238	−8	11	235	10
6	230	−5	12	245	13

(b) We see that
$$\begin{aligned} c_2 &= 300 + (-20) \\ c_3 &= 300 + (-20) + (-17) \\ c_4 &= 300 + (-20) + (-17) + (-14) \\ c_n &= 300 + \underbrace{(-20) + (-17) + (-14) + \cdots}_{n-1 \text{ terms}} \\ &= 300 + \sum_{i=1}^{n-1} a_i, \end{aligned}$$

where a_i is the ith term in the arithmetic sequence $-20, -17, -14, \ldots$, where $a_1 = -20$ and where $d = 3$. Using our formula for the sum of an arithmetic sequence, we know that

$$\begin{aligned}\sum_{i=1}^{n-1} a_i &= \frac{1}{2}(n-1)\left[2a_1 + (n-2)d\right] \qquad a_1 = -20 \text{ and } d = 3\\ &= \frac{1}{2}(n-1)\left[2(-20) + (n-2)(3)\right]\\ &= \frac{1}{2}(n-1)(3n-46).\end{aligned}$$

Thus, $c_n = 300 + \frac{1}{2}(n-1)(3n-46)$. To check this formula, we let $n = 12$:

$$c_{12} = 300 + \frac{1}{2}(12-1)(3\cdot 12 - 46) = 245,$$

which is the answer we got in part (a). To find the cost of producing the 50th piece of furniture, we have

$$c_{50} = 300 + \frac{1}{2}(50-1)(3\cdot 50 - 46) = 2848.$$

(c) From part (a), we see that producing the 8th piece costs only \$223, but that each subsequent piece costs more. From part (b), we see that producing the 50th piece costs \$2848—far more than is profitiable. Checking numbers, we see that the 19th and 20th pieces cost

$$\begin{aligned}c_{19} &= 300 + \frac{1}{2}(19-1)(3\cdot 19 - 46) = \$399\\ c_{20} &= 300 + \frac{1}{2}(20-1)(3\cdot 20 - 46) = \$433.\end{aligned}$$

We see that pieces 1 through 19 can be produced at a profit, although the profit for the 19th peice is only \$1. Pieces 20 and on are produced at a loss. Thus, the workshop should produce 19 pieces of furniture a day.

Solutions for Section 11.2

1. Yes, $a = 2$, ratio $= 1/2$.
5. Yes, $a = 1$, ratio $= -x$.
9. Sum $= \dfrac{1}{1-(-x)} = \dfrac{1}{1+x}$, $|x| < 1$.
13. After the initial term, each of these terms comes from multiplying 5 by the preceding number. So we know that we need powers of five to produce the correct series. One possibility is: $\sum_{n=0}^{6} 2(5^n)$.
17. $3 + \frac{3}{2} + \frac{3}{4} + \frac{3}{8} \cdots + \frac{3}{2^{10}} = 3\left(1 + \frac{1}{2} + \cdots + \frac{1}{2^{10}}\right) = \dfrac{3\left(1 - \frac{1}{2^{11}}\right)}{1 - \frac{1}{2}} = \dfrac{3\left(2^{11}-1\right)}{2^{10}} \approx 5.997.$
21. $$\sum_{i=4}^{\infty}\left(\frac{1}{3}\right)^i = \left(\frac{1}{3}\right)^4 + \left(\frac{1}{3}\right)^5 + \cdots = \left(\frac{1}{3}\right)^4\left(1 + \frac{1}{3} + \left(\frac{1}{3}\right)^2 + \cdots\right) = \frac{(\frac{1}{3})^4}{1-\frac{1}{3}} = \frac{1}{54}$$
25. (a) $0.232323\ldots = 0.23 + 0.23(0.01) + 0.23(0.01)^2 + \ldots$ which is a geometric series with $a = 0.23$ and $x = 0.01$.
    (b) The sum is $\dfrac{0.23}{1-0.01} = \dfrac{0.23}{0.99} = \dfrac{23}{99}$.
29. $0.4788888\ldots = 0.47 + \dfrac{8}{1{,}000} + \dfrac{8}{10{,}000} + \dfrac{8}{100{,}000} + \cdots$. Thus,
$$S = 0.47 + \frac{\frac{8}{1000}}{1 - \frac{1}{10}} = \frac{47}{100} + \frac{8}{900} = \frac{431}{900}.$$

33. Let Q_n represent the quantity, in milligrams, of ampicillin in the blood right after the nth tablet. Then

$$\begin{aligned} Q_1 &= 250 \\ Q_2 &= 250 + 250(0.04) \\ Q_3 &= 250 + 250(0.04) + 250(0.04)^2 \\ &\vdots \\ Q_n &= 250 + 250(1.04) + 250(1.04)^2 + \cdots + 250(0.04)^{n-1}. \end{aligned}$$

This is a geometric series. Its sum is given by

$$Q_n = \frac{250(1 - (0.04)^n)}{1 - 0.04}.$$

Thus,

$$Q_3 = \frac{250(1 - (0.04)^3)}{1 - 0.04} = 260.40$$

and

$$Q_{40} = \frac{250(1 - (0.04)^{40})}{1 - 0.04} = 260.417.$$

In the long run, as $n \to \infty$, we know that $(0.04)^n \to 0$, and so

$$Q_n = \frac{250(1 - (0.04)^n)}{1 - 0.04} \to \frac{250(1 - 0)}{1 - 0.04} = 260.417.$$

In the long run, the drug level approaches 260.417 mg.

37.

$$\begin{aligned} \text{Total present value, in dollars} &= 1000 + 1000e^{-0.04} + 1000e^{-0.04(2)} + 1000e^{-0.04(3)} + \cdots \\ &= 1000 + 1000(e^{-0.04}) + 1000(e^{-0.04})^2 + 1000(e^{-0.04})^3 + \cdots \end{aligned}$$

This is an infinite geometric series with $a = 1000$ and $x = e^{(-0.04)}$, and sum

$$\text{Total present value, in dollars} = \frac{1000}{1 - e^{-0.04}} = 25{,}503.$$

Solutions for Section 11.3

1. The graph of the parametric equations is in Figure 11.1.

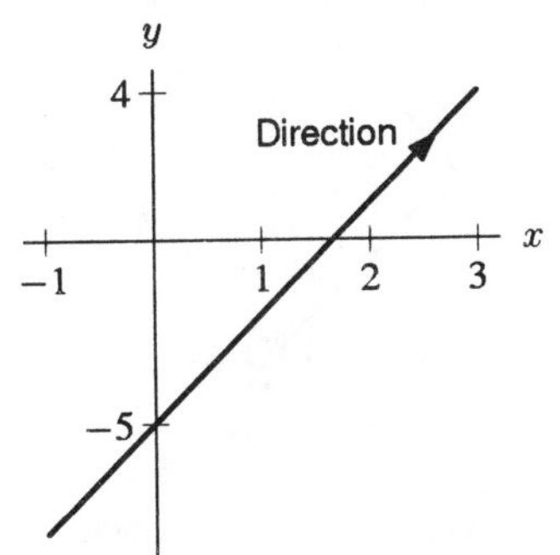

Figure 11.1

Since $x = t + 1$, we have $t = x - 1$. Substitute this into the second equation:

$$\begin{aligned} y &= 3t - 2 \\ y &= 3(x - 1) - 2 \\ y &= 3x - 5. \end{aligned}$$

5. The graph of the parametric equations is in Figure 11.2.

Since $x = t - 3$, we have $t = x + 3$. Substitute this into the second equation:

$$\begin{aligned} y &= t^2 + 2t + 1 \\ y &= (x+3)^2 + 2(x+3) + 1 \\ &= x^2 + 8x + 16 = (x+4)^2. \end{aligned}$$

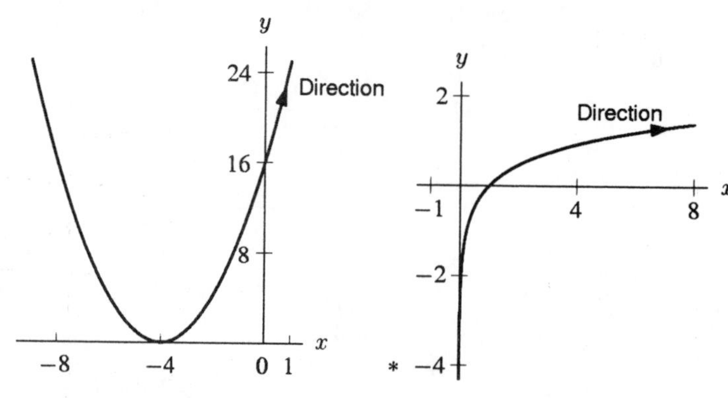

Figure 11.2 *Figure 11.3*

9. The graph of the parametric equations is in Figure 11.3.

Since $x = t^3$, we take the natural log of both sides and get $\ln x = 3\ln t$ or $\ln t = 1/3 \ln x$. We are given that $y = 2\ln t$, thus,

$$y = 2\left(\frac{1}{3}\ln x\right) = \frac{2}{3}\ln x.$$

13. Between times $t = 0$ and $t = 1$, x goes at a constant rate from 0 to 1 and y goes at a constant rate from 1 to 0. So the particle moves in a straight line from $(0, 1)$ to $(1, 0)$. Similarly, between times $t = 1$ and $t = 2$, it goes in a straight line to $(0, -1)$, then to $(-1, 0)$, then back to $(0, 1)$. So it traces out the diamond shown in Figure 11.4.

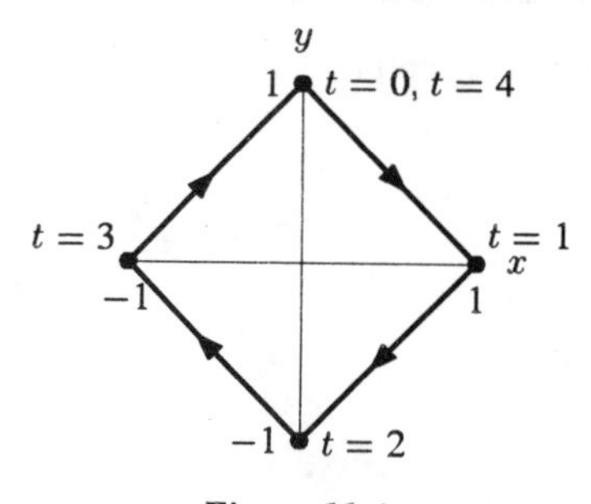

Figure 11.4

17. (a) We can replace x with t and $t + 1$ to get two parameterizations:

$$x = t, \quad y = t^2 \qquad \text{and} \qquad x = t + 1, \quad y = (t+1)^2$$

The second parameterization for x and y could look totally different:

$$x = t^3, \quad y = t^6.$$

(b) Altering the answers to part (a) gives:

$$x = t, \quad y = (t+2)^2 + 1 \qquad \text{and} \qquad x = t + 1, \quad y = (t+3)^2 + 1.$$

21. Let $f(t) = \ln t$. The particle is moving counterclockwise when $t > 0$. Any other time, when $t \leq 0$, the position is not defined.

25. (a) Since the x-coordinate and the y-coordinate are always the same (they both equal t), the bug follows the path $y = x$.
    (b) The bug starts at $(1, 0)$ because $\cos 0 = 1$ and $\sin 0 = 0$. Since the x-coordinate is $\cos x$, and the y-coordinate is $\sin x$, the bug follows the path of a unit circle, traveling counterclockwise. It reaches the starting point of $(1, 0)$ when $t = 2\pi$, because $\sin t$ and $\cos t$ are periodic with period 2π.
    (c) Now the x-coordinate varies from 1 to -1, while the y-coordinate varies from 2 to -2; otherwise, this is much like part (b) above. If we plot several points, the path looks like an ellipse, which is a circle stretched out in one direction.

29. For $0 \leq t \leq 2\pi$, the graph is in Figure 11.5.

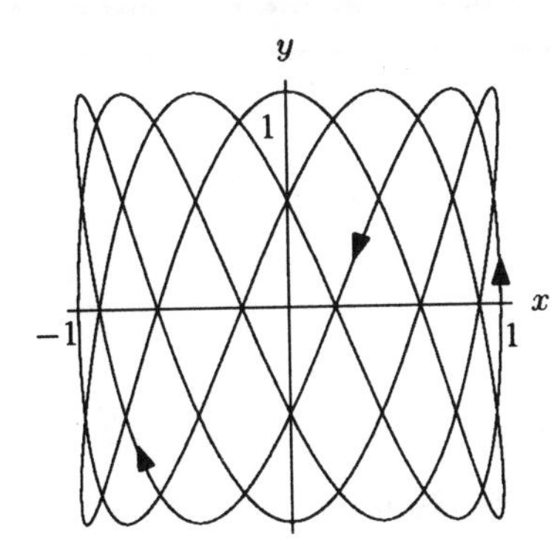

Figure 11.5

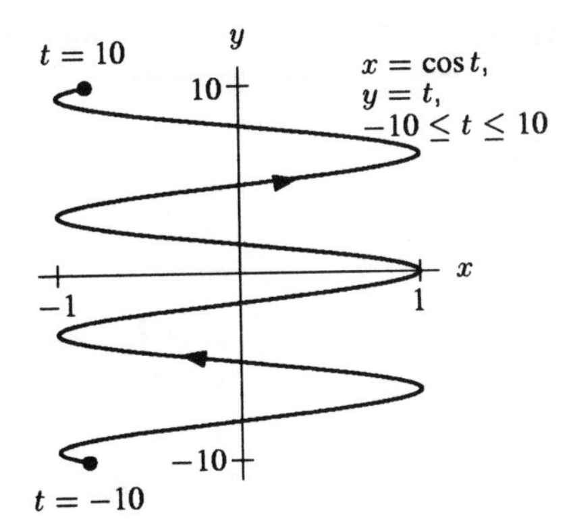

Figure 11.6

33. The particle moves back and forth between -1 and 1. See Figure 11.6.

Solutions for Section 11.4

1. (a) Center is $(2, -4)$ and radius is $\sqrt{20}$.
   (b) Rewriting the original equation, we have

$$\begin{aligned} 2x^2 + 2y^2 + 4x - 8y &= 12 \\ x^2 + y^2 + 2x - 4y &= 6 \\ (x^2 + 2x + 1) + (y^2 - 4y + 4) - 5 &= 6 \\ (x + 1)^2 + (y - 2)^2 &= 11. \end{aligned}$$

   So the center is $(-1, 2)$, and the radius is $\sqrt{11}$.

5. Group the terms and complete the square

$$\begin{aligned} x^2 + 4y^2 + 2x - 8y &= 11 \\ x^2 + 2x + 1 + 4(y^2 - 2y + 1) &= 11 + 1 + 4 \\ (x + 1)^2 + 4(y - 1)^2 &= 16 \end{aligned}$$

   Divide through by 16,

$$\frac{(x + 1)^2}{16} + \frac{(y - 1)^2}{4} = 1.$$

   This is an ellipse, centered at $(-1, 1)$, with horizontal axis of length 8 and vertical axis of length 4.

9. $(x - 2)^2 + (y - 3)^2 = 36$.

13. Since $2a = 4$ and $2b = 8$, giving $a = 2$, $b = 4$, so we have

$$\frac{(x + 4)^2}{4} + \frac{(y + 2)^2}{16} = 1.$$

17. The parameterization $(x, y) = (4+4\cos t, 4+4\sin t)$ gives the correct circle, but starts at $(8,4)$. To start on the x-axis we need $y = 0$ at $t = 0$, thus

$$(x, y) = \left(4+4\cos\left(t-\frac{\pi}{2}\right), 4+4\sin\left(t-\frac{\pi}{2}\right)\right).$$

21. The parabola $y = (x-2)^2$, for $1 \le x \le 3$.

25. Implicit: $x^2 - 2x + y^2 = 0$, $y < 0$. Explicit: $y = -\sqrt{-x^2+2x}$, $0 \le x \le 2$. Parametric: The curve is the lower half of a circle centered at $(1,0)$ with radius 1, so $x = 1+\cos t$, $y = \sin t$, for $\pi \le t \le 2\pi$.

Solutions for Section 11.5

1. $2e^{\frac{i\pi}{2}}$

5. $e^{\frac{i3\pi}{2}}$

9. $-5+12i$

13. We have $\sqrt{e^{i\pi/3}} = e^{(i\pi/3)/2} = e^{i\pi/6}$, thus $\cos\frac{\pi}{6} + i\sin\frac{\pi}{6} = \frac{\sqrt{3}}{2} + \frac{i}{2}$.

17. One value of $\sqrt[3]{i}$ is $\sqrt[3]{e^{i\frac{\pi}{2}}} = (e^{i\frac{\pi}{2}})^{\frac{1}{3}} = e^{i\frac{\pi}{6}} = \cos\frac{\pi}{6} + i\sin\frac{\pi}{6} = \frac{\sqrt{3}}{2} + \frac{i}{2}$

21. One value of $(\sqrt{3}+i)^{-1/2}$ is
$(2e^{i\frac{\pi}{6}})^{-1/2} = \frac{1}{\sqrt{2}}e^{i(-\frac{\pi}{12})} = \frac{1}{\sqrt{2}}\cos(-\frac{\pi}{12}) + i\frac{1}{\sqrt{2}}\sin(-\frac{\pi}{12}) \approx 0.683 - 0.183i$

25. (a) $z_1z_2 = (-3-i\sqrt{3})(-1+i\sqrt{3}) = 3+(\sqrt{3})^2 + i(\sqrt{3}-3\sqrt{3}) = 6 - i2\sqrt{3}$.

$$\frac{z_1}{z_2} = \frac{-3-i\sqrt{3}}{-1+i\sqrt{3}} \cdot \frac{-1-i\sqrt{3}}{-1-i\sqrt{3}} = \frac{3-(\sqrt{3})^2 + i(\sqrt{3}+3\sqrt{3})}{(-1)^2+(\sqrt{3})^2} = \frac{i\cdot 4\sqrt{3}}{4} = i\sqrt{3}.$$

(b) We find (r_1, θ_1) corresponding to $z_1 = -3 - i\sqrt{3}$.
$r_1 = \sqrt{(-3)^2 + (\sqrt{3})^2} = \sqrt{12} = 2\sqrt{3}$.
$\tan\theta_1 = \dfrac{-\sqrt{3}}{-3} = \dfrac{\sqrt{3}}{3}$, so $\theta_1 = \dfrac{7\pi}{6}$.
Thus $-3 - i\sqrt{3} = r_1e^{i\theta_1} = 2\sqrt{3}e^{i\frac{7\pi}{6}}$.

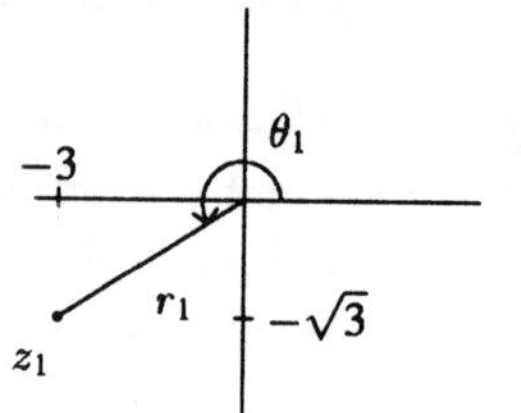

We find (r_2, θ_2) corresponding to $z_2 = -1 + i\sqrt{3}$. $r_2 = \sqrt{(-1)^2+(\sqrt{3})^2} = 2$;
$\tan\theta_2 = \dfrac{\sqrt{3}}{-1} = -\sqrt{3}$, so $\theta_2 = \dfrac{2\pi}{3}$.
Thus, $-1 + i\sqrt{3} = r_2e^{i\theta_2} = 2e^{i\frac{2\pi}{3}}$.

z_2 $\sqrt{3}$ r_2 θ_2 -1

We now calculate z_1z_2 and $\dfrac{z_1}{z_2}$.

$$z_1z_2 = \left(2\sqrt{3}e^{i\frac{7\pi}{6}}\right)\left(2e^{i\frac{2\pi}{3}}\right) = 4\sqrt{3}e^{i(\frac{7\pi}{6}+\frac{2\pi}{3})} = 4\sqrt{3}e^{i\frac{11\pi}{6}}$$
$$= 4\sqrt{3}\left[\cos\frac{11\pi}{6} + i\sin\frac{11\pi}{6}\right] = 4\sqrt{3}\left[\frac{\sqrt{3}}{2} - i\frac{1}{2}\right] = 6 - i2\sqrt{3}.$$

$$\frac{z_1}{z_2} = \frac{2\sqrt{3}e^{i\frac{7\pi}{6}}}{2e^{i\frac{2\pi}{3}}} = \sqrt{3}e^{i(\frac{7\pi}{6}-\frac{2\pi}{3})} = \sqrt{3}e^{i\frac{\pi}{2}}$$
$$= \sqrt{3}\left(\cos\frac{\pi}{2} + i\sin\frac{\pi}{2}\right) = i\sqrt{3}.$$

These agree with the values found in (a).

29. False, since $(1+i)^2 = 2i$ is not real.

33. Using Euler's formula, we have:

$$e^{i(2\theta)} = \cos 2\theta + i\sin 2\theta$$

On the other hand,

$$e^{i(2\theta)} = (e^{i\theta})^2 = (\cos\theta + i\sin\theta)^2 = (\cos^2\theta - \sin^2\theta) + i(2\cos\theta\sin\theta)$$

Equating imaginary parts, we find

$$\sin 2\theta = 2\sin\theta\cos\theta.$$

Solutions for Section 11.6

1. Substitute $x = 0$ into the formula for $\sinh x$. This yields

$$\sinh 0 = \frac{e^0 - e^{-0}}{2} = \frac{1-1}{2} = 0.$$

5. First, we observe that

$$\cosh 2x = \frac{e^{2x} + e^{-2x}}{2}.$$

Now, using the fact that $e^x \cdot e^{-x} = 1$, we calculate

$$\begin{aligned}\cosh^2 x &= \left(\frac{e^x + e^{-x}}{2}\right)^2 \\ &= \frac{(e^x)^2 + 2e^x \cdot e^{-x} + (e^{-x})^2}{4} \\ &= \frac{e^{2x} + 2 + e^{-2x}}{4}.\end{aligned}$$

Similarly, we have

$$\begin{aligned}\sinh^2 x &= \left(\frac{e^x - e^{-x}}{2}\right)^2 \\ &= \frac{(e^x)^2 - 2e^x \cdot e^{-x} + (e^{-x})^2}{4} \\ &= \frac{e^{2x} - 2 + e^{-2x}}{4}.\end{aligned}$$

Thus, to obtain $\cosh 2x$, we need to add (rather than subtract) $\cosh^2 x$ and $\sinh^2 x$, giving

$$\begin{aligned}\cosh^2 x + \sinh^2 x &= \frac{e^{2x} + 2 + e^{-2x} + e^{2x} - 2 + e^{-2x}}{4} \\ &= \frac{2e^{2x} + 2e^{-2x}}{4} \\ &= \frac{e^{2x} + e^{-2x}}{2} \\ &= \cosh 2x.\end{aligned}$$

Thus, we see that the identity relating $\cosh 2x$ to $\cosh x$ and $\sinh x$ is

$$\cosh 2x = \cosh^2 x + \sinh^2 x.$$

9. We know that $\sinh(iz) = i\sin z$, where z is real. Substituting $z = ix$, where x is real so z is imaginary, we have

$$\begin{aligned} \sinh(iz) &= i\sin z \\ \sinh(i \cdot ix) &= i\sin(ix) \qquad \text{substituting } z = ix \\ \sinh(-x) &= i\sin(ix). \end{aligned}$$

But $\sinh(-x) = -\sinh(x)$, thus we have

$$-\sinh x = i\sin(ix).$$

Multiplying both sides by i gives

$$-i\sinh x = -1\sin(ix).$$

Thus,

$$i\sinh x = \sin(ix).$$

Solutions for Chapter 11 Review

1. We start with 100, decrease by tens until we reach 0, which is $100 - 10(10)$. A possible answer is

$$\sum_{n=0}^{10}(100 - 10n).$$

5. Yes, $a = 1$, ratio $= 2z$.

9. (a) Let h_n be the height of the nth bounce after the ball hits the floor for the nth time. Then from Figure 11.7,

$$\begin{aligned} h_0 &= \text{height before first bounce} = 10 \text{ feet}, \\ h_1 &= \text{height after first bounce} = 10\left(\frac{3}{4}\right) \text{ feet}, \\ h_2 &= \text{height after second bounce} = 10\left(\frac{3}{4}\right)^2 \text{ feet}. \end{aligned}$$

Generalizing this gives

$$h_n = 10\left(\frac{3}{4}\right)^n.$$

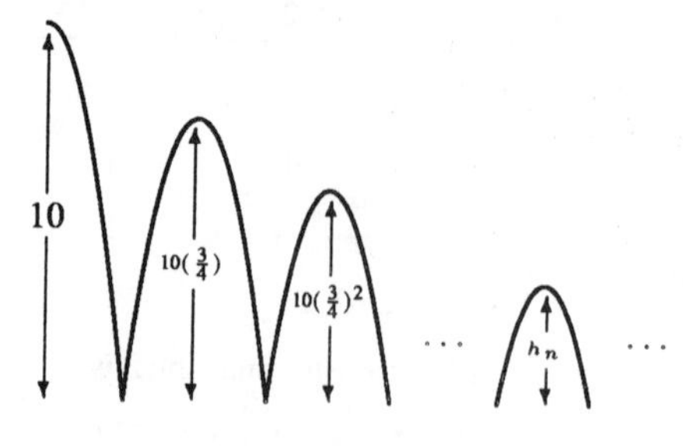

Figure 11.7

(b) When the ball hits the floor for the first time, the total distance it has traveled is just $D_1 = 10$ feet. (Notice that this is the same as $h_0 = 10$.) Then the ball bounces back to a height of $h_1 = 10\left(\frac{3}{4}\right)$, comes down and hits the floor for the second time. The total distance it has traveled is

$$D_2 = h_0 + 2h_1 = 10 + 2 \cdot 10\left(\frac{3}{4}\right) = 25 \text{ feet.}$$

Then the ball bounces back to a height of $h_2 = 10\left(\frac{3}{4}\right)^2$, comes down and hits the floor for the third time. It has traveled

$$D_3 = h_0 + 2h_1 + 2h_2 = 10 + 2 \cdot 10\left(\frac{3}{4}\right) + 2 \cdot 10\left(\frac{3}{4}\right)^2 = 25 + 2 \cdot 10\left(\frac{3}{4}\right)^2 = 36.25 \text{ feet.}$$

Similarly,

$$\begin{aligned} D_4 &= h_0 + 2h_1 + 2h_2 + 2h_3 \\ &= 10 + 2 \cdot 10\left(\frac{3}{4}\right) + 2 \cdot 10\left(\frac{3}{4}\right)^2 + 2 \cdot 10\left(\frac{3}{4}\right)^3 \\ &= 36.25 + 2 \cdot 10\left(\frac{3}{4}\right)^3 \\ &\approx 44.69 \text{ feet.} \end{aligned}$$

(c) When the ball hits the floor for the nth time, its last bounce was of height h_{n-1}. Thus, by the method used in part (b), we get

$$\begin{aligned} D_n &= h_0 + 2h_1 + 2h_2 + 2h_3 + \cdots + 2h_{n-1} \\ &= 10 + \underbrace{2 \cdot 10\left(\frac{3}{4}\right) + 2 \cdot 10\left(\frac{3}{4}\right)^2 + 2 \cdot 10\left(\frac{3}{4}\right)^3 + \cdots + 2 \cdot 10\left(\frac{3}{4}\right)^{n-1}}_{\text{finite geometric series}} \\ &= 10 + 2 \cdot 10 \cdot \left(\frac{3}{4}\right)\left(1 + \left(\frac{3}{4}\right) + \left(\frac{3}{4}\right)^2 + \cdots + \left(\frac{3}{4}\right)^{n-2}\right) \\ &= 10 + 15\left(\frac{1 - \left(\frac{3}{4}\right)^{n-1}}{1 - \left(\frac{3}{4}\right)}\right) \\ &= 10 + 60\left(1 - \left(\frac{3}{4}\right)^{n-1}\right). \end{aligned}$$

13. $x = t, y = 5$.

17. The plot looks like Figure 11.8.

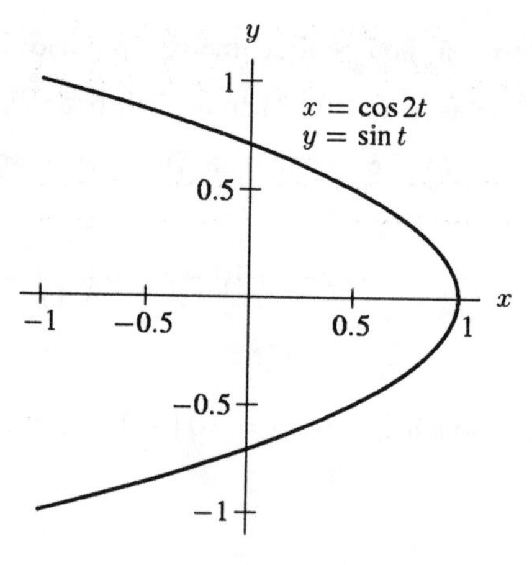

Figure 11.8

which does appear to be part of a parabola. To prove that it is, we note that we have

$$x = \cos 2t$$

$$y = \sin t$$

and must somehow find a relationship between x and y. Recall the trigonometric identity

$$\cos 2t = 1 - 2\sin^2 t.$$

Thus we have $x = 1 - 2y^2$, which is a parabola lying along the x-axis, for $-1 \le y \le 1$.

21. $13e^{i\theta}$, where $\theta = \arctan(-\frac{12}{5}) \approx -1.176$ is an angle in the fourth quadrant.

25. One value of $(-4+4i)^{2/3}$ is $[\sqrt{32}e^{i\frac{3\pi}{4}}]^{2/3} = (\sqrt{32})^{2/3}e^{i\frac{\pi}{2}} = 2^{\frac{10}{3}}\cos\frac{\pi}{2} + i2^{\frac{10}{3}}\sin\frac{\pi}{2} = 8i\sqrt[3]{2}$

APPENDIX

Solutions for Section A

1. $4^3 = 4 \cdot 4 \cdot 4 = 64$
5. $(-1)^{12} = \underbrace{(-1)(-1)\cdots(-1)}_{\text{12 factors}} = 1$
9. $\frac{10^8}{10^5} = 10^{8-5} = 10^3 = 10 \cdot 10 \cdot 10 = 1{,}000$
13. $\sqrt{4^2} = 4$
17. $\sqrt{x^4} = (x^4)^{1/2} = x^{4/2} = x^2$
21. $\sqrt{16x^3} = (16x^3)^{1/2} = 16^{1/2}x^{3/2} = 4x^{3/2}$
25. $\sqrt{r^3} = (r^3)^{1/2} = r^{3/2}$
29.
$$\begin{aligned}\sqrt{50x^4y^6} &= 50^{1/2} \cdot (x^4)^{1/2} \cdot (y^6)^{1/2} \\ &= 50^{1/2}x^2y^3 \\ &= (25 \cdot 2)^{1/2}x^2y^3 \\ &= 25^{1/2} \cdot 2^{1/2} \cdot x^2 \cdot y^3 \\ &= 5\sqrt{2}x^2y^3\end{aligned}$$

33. $(32)^{1/5} = (2^5)^{1/5} = 2^1 = 2$
37. $16^{5/4} = (2^4)^{5/4} = 2^5 = 32$
41. $3^{-(3/2)} = \frac{1}{3^{3/2}} = \frac{1}{(3^3)^{1/2}} = \frac{1}{(27)^{1/2}} = \frac{1}{(9 \cdot 3)^{1/2}} = \frac{1}{9^{1/2} \cdot 3^{1/2}} = \frac{1}{3\sqrt{3}}$
45. Since $\frac{1}{7^{-2}}$ is the same as 7^2, we obtain $7 \cdot 7$ or 49.
49. The order of operations tells us to square 3 first (giving 9) and then multiply by -2. Therefore $(-2)\left(3^2\right) = (-2)(9) = -18$.
53. First we see within the radical that $(-4)^2 = 16$. Therefore $\sqrt{(-4)^2} = \sqrt{16} = 4$.
57. For this example, we have $\left(\frac{1}{27}\right)^{-1/3} = (27)^{1/3} = 3$. This is because $\left(\frac{1}{27}\right)^{-1/3} = \left(\left(\frac{1}{27}\right)^{-1}\right)^{1/3} = \left(\frac{27}{1}\right)^{1/3} = 3$.
61. First we raise $3^{x/2}$ to the second power and multiply this result by 3. Therefore $3\left(3^{x/2}\right)^2 = 3(3^x) = 3^1(3^x) = 3^{x+1}$.
65. $\sqrt{e^{2x}} = (e^{2x})^{\frac{1}{2}} = e^{2x \cdot \frac{1}{2}} = e^x$
69. Inside the parenthesis we write the radical as an exponent, which results in
$$\left(3x\sqrt{x^3}\right)^2 = \left(3x \cdot x^{3/2}\right)^2.$$
Then within the parenthesis we write
$$\left(3x^1 \cdot x^{3/2}\right)^2 = \left(3x^{5/2}\right)^2 = 3^2(x^{5/2})^2 = 9x^5.$$

73. $\dfrac{4A^{-3}}{(2A)^{-4}} = \dfrac{4/A^3}{1/(2A)^4} = \dfrac{4}{A^3} \cdot \dfrac{(2A)^4}{1} = \dfrac{4}{A^3} \cdot \dfrac{2^4A^4}{1} = 64A.$

77. First we divide within the larger parentheses. Therefore,

$$\left(\frac{35(2b+1)^9}{7(2b+1)^{-1}}\right)^2 = \left(5(2b+1)^{9-(-1)}\right)^2 = \left(5(2b+1)^{10}\right)^2.$$

Then we expand to obtain

$$25(2b+1)^{20}.$$

81. $(-625)^{3/4} = (\sqrt[4]{-625})^3$. Since $\sqrt[4]{-625}$ is not a real number, $(-625)^{3/4}$ is undefined.

85. $(-64)^{3/2} = (\sqrt{-64})^3$. Since $\sqrt{-64}$ is not a real number, $(-64)^{3/2}$ is undefined.

89. True

93. True

Solutions for Section B

1. $-(x-3) - 2(5-x) = -x + 3 - 10 + 2x = x - 7.$

5. $2(3x-7) = 6x - 14$

9. $x(2x+5) = 2x^2 + 5x$

13. $5z(x-2) - 3(x-2) = 5xz - 10z - 3x + 6$

17. $(x+2)(3x-8) = 3x^2 - 8x + 6x - 16 = 3x^2 - 2x - 16$

21. First we multiply 4 by the terms $3x$ and $-2x^2$, and then use foil to expand $(5+4x)(3x-4)$. Therefore,

$$\begin{aligned}\left(3x-2x^2\right)(4) + (5+4x)(3x-4) &= 12x - 8x^2 + 15x - 20 + 12x^2 - 16x \\ &= 4x^2 + 11x - 20.\end{aligned}$$

25. The order of operations tells us to expand $(x-3)^2$ first and then multiply the result by 4. Therefore,

$$\begin{aligned}4(x-3)^2 + 7 &= 4(x-3)(x-3) + 7 \\ &= 4(x^2 - 3x - 3x + 9) + 7 = 4(x^2 - 6x + 9) + 7 \\ &= 4x^2 - 24x + 36 + 7 = 4x^2 - 24x + 43.\end{aligned}$$

29. Using foil we obtain:

$$\begin{aligned}(x+3)\left(\frac{24}{x} + 2\right) &= 24 + 2x + \frac{72}{x} + 6 \\ &= 30 + 2x + \frac{72}{x}.\end{aligned}$$

Solutions for Section C

1. $2x + 6 = 2(x+3)$

5. $10w - 25 = 5(2w - 5)$

9. $12x^3y^2 - 18x = 6x(2x^2y^2 - 3)$

13. $x^2 - 3x + 2 = (x-2)(x-1)$

17. Can be factored no further.

21. Since each term has a common factor of 2, we write:

$$2x^2 - 10x + 12 = 2\left(x^2 - 5x + 6\right) = 2(x-3)(x-2).$$

25. $x^2 - 1.4x - 3.92 = (x + 1.4)(x - 2.8)$

29. $c^2 + x^2 - 2cx = (x - c)^2$

33. By grouping the terms hx^2 and $-4hx$, we find a common factor of hx and for the terms 12 and $-3x$, we find a common factor of -3. Therefore,

$$hx^2 + 12 - 4hx - 3x = hx^2 - 4hx + 12 - 3x = hx(x-4) - 3(-4 + x) = hx(x-4) - 3(x-4) = (hx-3)(x-4).$$

37. $t^2e^{5t} + 3te^{5t} + 2e^{5t} = e^{5t}(t^2 + 3t + 2) = e^{5t}(t+1)(t+2)$.

41. The quadratic expression in this expression factors into two binomials as:

$$\left(\cos^2 x - 2\cos x + 1\right) = (\cos x - 1)(\cos x - 1) = (\cos x - 1)^2.$$

Solutions for Section D

1. $\frac{3}{5} + \frac{4}{7} = \frac{3 \cdot 7 + 4 \cdot 5}{35} = \frac{21 + 20}{35} = \frac{41}{35}$

5. $\frac{-2}{yz} + \frac{4}{z} = \frac{-2z + 4yz}{yz^2} = \frac{-2 + 4y}{yz} = \frac{-2(1-2y)}{yz}$

9. $\frac{\frac{3}{4}}{\frac{7}{20}} = \frac{3}{4} \cdot \frac{20}{7} = \frac{60}{28} = \frac{15}{7}$

13. $\frac{13}{x-1} + \frac{14}{2x-2} = \frac{13}{x-1} + \frac{14}{2(x-1)} = \frac{13 \cdot 2 + 14}{2(x-1)} = \frac{40}{2(x-1)} = \frac{20}{x-1}$

17. $\frac{8y}{y-4} + \frac{32}{y-4} = \frac{8y+32}{y-4} = \frac{8(y+4)}{y-4}$

21.

$$\frac{8}{3x^2 - x - 4} - \frac{9}{x+1} = \frac{8}{(x+1)(3x-4)} - \frac{9}{x+1} = \frac{8 - 9(3x-4)}{(x+1)(3x-4)} = \frac{-27x + 44}{(x+1)(3x-4)}$$

25. If we rewrite the second fraction $-\frac{1}{1-x}$ as $\frac{1}{x-1}$, the common denominator becomes $x - 1$. Therefore,

$$\frac{x^2}{x-1} - \frac{1}{1-x} = \frac{x^2}{x-1} + \frac{1}{x-1} = \frac{x^2+1}{x-1}.$$

29. The common denominator is e^{2x}. Thus,

$$\frac{1}{e^{2x}} + \frac{1}{e^x} = \frac{1}{e^{2x}} + \frac{e^x}{e^{2x}} = \frac{1 + e^x}{e^{2x}}.$$

33. $\frac{8y}{y-4} - \frac{32}{y-4} = \frac{8y - 32}{y-4} = \frac{8(y-4)}{y-4} = 8$

37. We write this complex fraction as a multiplication problem. Therefore,

$$\frac{\frac{w+2}{2}}{w+2} = \frac{w+2}{2} \cdot \frac{1}{w+2} = \frac{1}{2}.$$

41. We expand within the first brackets first. Therefore,

$$\frac{[4-(x+h)^2]-[4-x^2]}{h} = \frac{[4-(x^2+2xh+h^2)]-[4-x^2]}{h}$$
$$= \frac{[4-x^2-2xh-h^2]-4+x^2}{h} = \frac{-2xh-h^2}{h}$$
$$= -2x-h.$$

45. We simplify the second complex fraction first. Thus,

$$p - \frac{q}{\frac{p}{q}+\frac{q}{p}} = p - \frac{q}{\frac{p^2+q^2}{qp}} = p - q \cdot \frac{qp}{p^2+q^2}$$
$$= \frac{p(p^2+q^2)-q^2p}{p^2+q^2} = \frac{p^3}{p^2+q^2}.$$

49. Write

$$\frac{\frac{1}{2}(2x-1)^{-1/2}(2)-(2x-1)^{1/2}(2x)}{(x^2)^2} = \frac{\frac{1}{(2x-1)^{1/2}} - \frac{2x(2x-1)^{1/2}}{1}}{(x^2)^2}.$$

Next a common denominator for the top two fractions is $(2x-1)^{1/2}$. Therefore we obtain,

$$\frac{\frac{1}{(2x-1)^{1/2}} - \frac{2x(2x-1)}{(2x-1)^{1/2}}}{x^4} = \frac{1-4x^2+2x}{(2x-1)^{1/2}} \cdot \frac{1}{x^4} = \frac{-4x^2+2x+1}{x^4\sqrt{2x-1}}.$$

53. The denominator p^2+11 is divided into each of the two terms of the numerator. Thus,

$$\frac{7+p}{p^2+11} = \frac{7}{p^2+11} + \frac{p}{p^2+11}.$$

57. The numerator $q-1 = q-4+3$. Thus,

$$\frac{q-1}{q-4} = \frac{(q-4)+3}{q-4} = 1 + \frac{3}{q-4}.$$

61.

$$\frac{1+e^x}{e^x} = \frac{1}{e^x} + \frac{e^x}{e^x} = \frac{1}{e^x} + 1 = 1 + \frac{1}{e^x}$$

65. False

Solutions for Section E

1. $3x^2\left(x^{-1}\right) + \frac{1}{2x} + x^2 + \frac{1}{5} = 3x^1 + \frac{1}{2}x^{-1} + x^2 + \frac{1}{5}$

5.

$$2P^2(P) + (9P)^{1/2} = 2P^3 + 3P^{1/2}$$

9.

$$\frac{-3(4x-x^2)}{7x} = \frac{-12x+3x^2}{7x} = -\frac{12}{7} + \frac{3}{7}x$$

13. We write the denominator as a factor in the numerator as:

$$\frac{12}{\sqrt{3x+1}} = \frac{12}{(3x+1)^{1/2}} = 12(3x+1)^{-1/2}.$$

17.

$$4(6R+2)^3(6) = 4(6)(6R+2)^3 = 24(6R+2)^3$$

21. $10,000(1-0.24)^t = 10,000(.76)^t$

25. Since $16^{t/2}$ represents the square root of 16 raised to the power of t, we have

$$16^{t/2} = \left(\sqrt{16}\right)^t = 4^t.$$

29. $\left(10e^t\right)^2 = 10^2 e^{2t} = 100(e^2)^t$.

33. $(2y^2-5)3y = 6y^3 - 15y$

37. $x^2+8x = x^2+8x+16-16 = (x+4)^2-16$

41. $s^2+6s-8 = s^2+6s+9-9-8 = (s+3)^2-17$

45. First we factor out -1. Then

$$\begin{aligned}-x^2+6x-2 &= -(x^2-6x+2) = -(x^2-6x+9-9+2)\\ &= -(x^2-6x+9-7) = -(x^2-6x+9)+7\\ &= -(x-3)^2+7.\end{aligned}$$

49. We write

$$-(\sin(\pi t))^{-1}(-\cos(\pi t))\pi = \frac{-(-\cos(\pi t))\pi}{\sin(\pi t)} = \frac{\pi\cos(\pi t)}{\sin(\pi t)}.$$

53. $(5x)^{1/2} = \sqrt{5x}$

57.

$$\begin{aligned}\frac{1}{2}(x^2+16)^{-1/2}(2x) &= \frac{1}{2}(2x)(x^2+16)^{-1/2} = x(x^2+16)^{-1/2}\\ &= \frac{x}{(x^2+16)^{1/2}} = \frac{x}{\sqrt{x^2+16}}\end{aligned}$$

Solutions for Section F

1.

$$\begin{aligned}3x &= 15\\ \frac{3x}{3} &= \frac{15}{3}\\ x &= 5\end{aligned}$$

5.

$$\begin{aligned}y-5 &= 21\\ +5 &= +5\\ y &= 26\end{aligned}$$

9.

$$\begin{aligned}13t+2 &= 47\\ -2 &= -2\\ 13t &= 45\\ \frac{13t}{13} &= \frac{45}{13}\\ t &= \frac{45}{13}\end{aligned}$$

13.

$$\begin{aligned} y^2 - 5y - 6 &= 0 \\ (y+1)(y-6) &= 0 \\ y+1 = 0 \quad &\text{or} \quad y-6 = 0 \\ y = -1 \quad &\text{or} \quad y = 6 \end{aligned}$$

17.

$$\begin{aligned} \frac{3}{x-1} + 1 &= 5 \\ \frac{3}{x-1} &= 4 \\ 4(x-1) &= 3 \\ 4x - 4 &= 3 \\ 4x &= 7 \\ x &= \frac{7}{4} \end{aligned}$$

21.

$$\begin{aligned} \frac{21}{z-5} - \frac{13}{z^2-5z} &= 3 \\ \frac{21}{z-5} - \frac{13}{z(z-5)} &= 3 \\ \frac{21z-13}{z(z-5)} &= 3 \\ 21z - 13 &= 3z(z-5) \\ 21z - 13 &= 3z^2 - 15z \\ 3z^2 - 36z + 13 &= 0 \end{aligned}$$

$$\begin{aligned} z &= \frac{-(-36) \pm \sqrt{(-36)^2 - 4(3)(13)}}{2(3)} \\ &= \frac{36 \pm \sqrt{1140}}{6} \\ &= \frac{36 \pm \sqrt{4 \cdot 285}}{6} \\ &= \frac{36 \pm 2\sqrt{285}}{6} \\ &= \frac{18 \pm \sqrt{285}}{3} \end{aligned}$$

25. $\frac{C}{2\pi} = r$

29.

$$\begin{aligned} 0.079 &= \ln B \\ e^{0.079} &= e^{\ln B} \\ e^{0.079} &= B \end{aligned}$$

33. First solve for b^5, then take fifth root:

$$\begin{aligned} Ab^5 &= C \\ b^5 &= \frac{C}{A} \\ b &= \sqrt[5]{\frac{C}{A}}. \end{aligned}$$

37. Expanding yields

$$\begin{aligned}1.06s - 0.01(248.4 - s) &= 22.67s\\ 1.06s - 2.484 + 0.01s &= 22.67s\\ -21.6s &= 2.484\\ s &= -0.115.\end{aligned}$$

41. By grouping the first two and the last two terms, we obtain:

$$\begin{aligned}\left(2p^3 + p^2\right) - 18p - 9 &= 0\\ \left(2p^3 + p^2\right) - (18p + 9) &= 0\\ p^2(2p+1) - 9(2p+1) &= 0\\ \left(p^2 - 9\right)(2p+1) &= 0\\ (p-3)(p+3)(2p+1) &= 0\end{aligned}$$

$$p = 3, \text{ or } p = -3, \text{ or } p = -\frac{1}{2}.$$

45.

$$\begin{aligned}4x^2 - 13x - 12 &= 0\\ (x-4)(4x+3) &= 0\\ x = 4 \text{ or } x &= -\frac{3}{4}\end{aligned}$$

49. First we combine like terms in the numerator.

$$\begin{aligned}\frac{x^2 + 1 - 2x^2}{(x^2+1)^2} &= 0\\ \frac{-x^2+1}{(x^2+1)^2} &= 0\\ -x^2 + 1 &= 0\\ -x^2 &= -1\\ x^2 &= 1\\ x &= \pm 1\end{aligned}$$

53. We can solve this equation by cubing both sides of this equation.

$$\begin{aligned}\frac{1}{\sqrt[3]{x}} &= -2\\ \left(\frac{1}{\sqrt[3]{x}}\right)^3 &= (-2)^3\\ \frac{1}{x} &= -8\\ x &= -\frac{1}{8}\end{aligned}$$

57. Use the fact that $5^{2x} = (5^x)^2$ and factor

$$\begin{aligned}5^{2x} - 5^x - 6 &= 0\\ (5^x + 2)(5^x - 3) &= 0\\ 5^x = -2 \text{ or } 5^x &= 3.\end{aligned}$$

$5^x = -2$ is impossible, so $5^x = 3$. Solve with graphing calculator, getting $x \approx 0.6826$.

61.

$$\frac{1}{2}\left(2^x\right) = 16$$
$$2^x = 32$$

Since $2^5 = 32$,

$$x = 5.$$

65. We begin by squaring both sides of the equation in order to eliminate the radical.

$$T = 2\pi\sqrt{\frac{l}{g}}$$
$$T^2 = 4\pi^2\left(\frac{l}{g}\right)$$
$$\frac{gT^2}{4\pi^2} = l$$

69. We collect all terms involving y and then factor out the y.

$$by - d = ay + c$$
$$by - ay = c + d$$
$$y(b - a) = c + d$$
$$y = \frac{c + d}{b - a}$$

73. Multiplying by N/MA:

$$\frac{NK}{MA} = y - z$$
$$z = y - \frac{NK}{MA}.$$

77. Before completing the square, get all terms on the left, and divide by the coefficient of x^2:

$$8x^2 - 1 = 2x$$
$$8x^2 - 2x - 1 = 0$$
$$x^2 - \frac{1}{4}x - \frac{1}{8} = 0.$$

Now complete the square

$$x^2 - \frac{1}{4}x + \frac{1}{64} - \frac{1}{64} - \frac{1}{8} = 0$$
$$(x - \frac{1}{8})^2 - \frac{9}{64} = 0$$
$$x - \frac{1}{8} = \pm\sqrt{\frac{9}{64}}$$
$$x = \frac{1}{8} \pm \frac{3}{8}$$

So the solutions are $x = 1/2$ and $x = -1/4$.

Solutions for Section G

1.
$$2x = 8$$
$$x = 4$$

Therefore,

$$4 + y = 3,$$

so

$$y = -1.$$

5. Substituting the value of y from the second equation into the first equation, we obtain

$$\begin{aligned} x^2 + (x-3)^2 &= 36 \\ x^2 + x^2 - 6x + 9 &= 36 \\ 2x^2 - 6x - 27 &= 0, \end{aligned}$$

$$\begin{aligned} x &= \frac{-(-6) \pm \sqrt{(-6)^2 - 4(2)(-27)}}{(2)(2)} \\ &= \frac{6 \pm \sqrt{252}}{4} \\ &= \frac{6 \pm \sqrt{4 \cdot 63}}{4} \\ &= \frac{6 \pm 2\sqrt{63}}{4} \\ &= \frac{3 \pm \sqrt{63}}{2}. \end{aligned}$$

Now we substitute the values of x into the second equation:

$$\begin{aligned} y &= \frac{3 \pm \sqrt{63}}{2} - 3 \\ &= \frac{-3 \pm \sqrt{63}}{2}. \end{aligned}$$

9. We set the equations $y = \frac{1}{x}$ and $y = 4x$ equal to one another.

$$\begin{aligned} \frac{1}{x} &= 4x \\ 4x^2 &= 1 \\ x^2 &= \frac{1}{4} \\ x &= \frac{1}{2} \quad \text{and} \quad y = \frac{1}{\frac{1}{2}} = 2 \quad \text{or} \\ x &= -\frac{1}{2} \quad \text{and} \quad y = \frac{1}{-\frac{1}{2}} = -2 \end{aligned}$$

13. Using the point-slope formula for the equation of a line, we have

$$y - 0 = 3(x - 0)$$

or

$$y = 3x.$$

We need to find the points where this line intersects $y = x^2$. This means we want points such that

$$x^2 = 3x \quad \text{or} \quad x^2 - 3x = 0$$
$$x(x-3) = 0$$
$$x = 0 \quad \text{or} \quad x = 3.$$

So the points are $(0, 0)$ and $(3, 9)$.

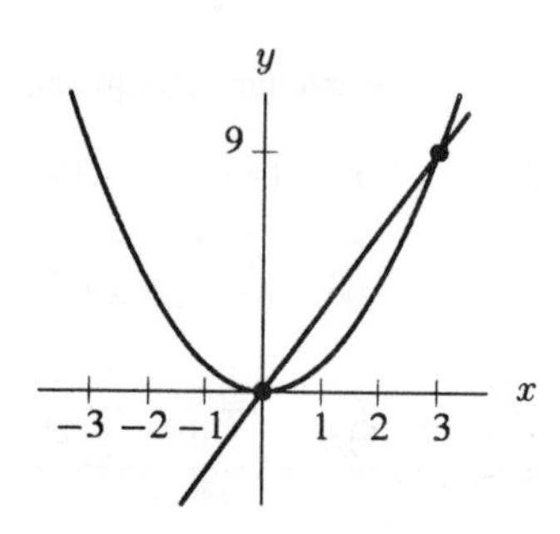

Figure G.1

Solutions for Section H

1.

$$2(x-7) \geq 0$$
$$2x - 14 \geq 0$$
$$2x \geq 14$$
$$x \geq 7$$

5. The expression $x + 4$ must be greater than or equal to zero. Therefore

$$x + 4 \geq 0$$
$$x \geq -4.$$

9.

$$5 - x < 0$$
$$5 < x$$
$$x > 5$$

13. Since $x^2 + 1$ is always positive, $\dfrac{x-5}{x^2+1} > 0$ if $x - 5 > 0$, so $x > 5$.

17. $-1 < y < 1$

21. $t \geq 1995$

25.

$$
\begin{aligned}
-4x + 7 &< 13 \\
-4x &< 6 \\
x &> -\frac{6}{4}
\end{aligned}
$$

−6/4 0

Figure H.2

29.

$$
\begin{aligned}
\frac{2}{x-5} &> 3 \\
\frac{2}{x-5} - 3 &> 0 \\
\frac{2}{x-5} - \frac{3(x-5)}{x-5} &> 0 \\
\frac{2 - 3x + 15}{x-5} &> 0 \\
\frac{-3x + 17}{x-5} &> 0
\end{aligned}
$$

We want the numerator and denominator to have the same sign. Also, $x \neq 5$. The numerator is positive when $-3x+17 > 0$ or $x < \frac{17}{3}$. The denominator is positive when $x > 5$, so $5 < x < \frac{17}{3}$ is a solution. The numerator and denominator are both negative when $-3x + 17 < 0$ or $x > \frac{17}{3}$ and $x < 5$. There are no xs that satisfy this inequality.

Figure H.3

33. $|2z - 7| < 8$ means

$$
\begin{aligned}
-8 < 2z - 7 &< 8 \\
-1 < 2z &< 15 \\
-\frac{1}{2} < z &< \frac{15}{2}.
\end{aligned}
$$

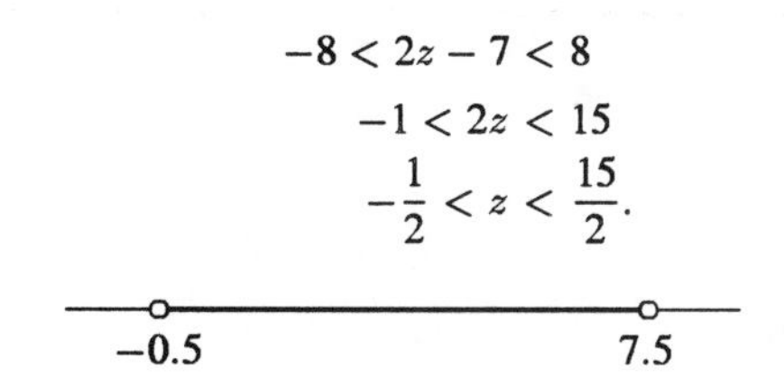

Figure H.4

37. (a) No value of x makes the expression undefined.

(b)

$$
\begin{aligned}
(2x)e^x + x^2e^x &= 0 \\
xe^x(2 + x) &= 0.
\end{aligned}
$$

Hence,

$$x = 0 \text{ or } x = -2.$$

(c) To solve $(2x)e^x + x^2e^x > 0$, or $xe^x(2 + x) > 0$, flag the number line at $x = -2, x = 0$.

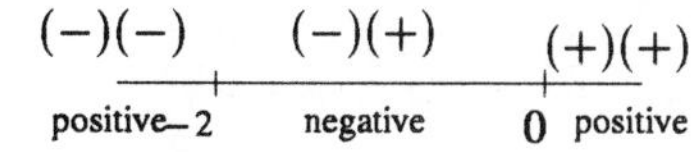

(−)(−) (−)(+) (+)(+)

positive −2 negative 0 positive

So $x > 0$ or $x < -2$.

(d) $(2x)e^x + x^2e^x < 0$ when $-2 < x < 0$.

41. (a) The fraction $-\frac{24}{p^3}$ is undefined at $p = 0$.

(b) Since no number when divided into -24 will yield zero, there is no solution to $\frac{-24}{p^3} = 0$.

(c) If $p < 0$, then $p^3 < 0$, and if $p^3 < 0$, then $\frac{-24}{p^3} > 0$. Thus, $\frac{-24}{p^3} > 0$ for $p < 0$.

(d) If $p > 0$, then $p^3 > 0$, and $\frac{-24}{p^3} < 0$. Thus, $\frac{-24}{p^3} < 0$ for $p > 0$.

45. (a) The expression is undefined when $x - 2 = 0$, i.e. when $x = 2$.

(b) The expression is zero when $x - 1 = 0$, i.e. when $x = 1$.

(c) Notice that $(x - 1) > 0$ when $x > 1$ and $(x - 2) > 0$ when $x > 2$. The expression is positive when both numerator and denominator are positive — that is, $x > 2$ — or when both numerator and denominator are negative — that is, $x < 1$.

(d) The expression is negative when $1 < x < 2$.

49. First we add 3 to all the terms of the inequality.

$$-1 \le 4x - 3 \le 1$$
$$2 \le 4x \le 4$$

Now divide by 4,

$$\frac{1}{2} \le x \le 1.$$

53. Factor by grouping produces

$$2(x-1)(x+4) + (x-1)^2 > 0$$
$$(x-1)(2(x+4) + (x-1)) > 0$$
$$(x-1)(2x+8+x-1) > 0$$
$$(x-1)(3x+7) > 0.$$

The critical values are

$$x = 1,\ -\frac{7}{3}.$$

Thus,

$(-)(-)$		$(-)(+)$		$(+)(+)$
positive	$-\frac{7}{3}$	negative	1	positive

Thus the solution is

$$x < -\frac{7}{3} \quad \text{or } x > 1.$$

57.

$$\frac{1}{x} > \frac{1}{x+1}$$
$$\frac{1}{x} - \frac{1}{x+1} > 0$$

Using a common denominator of $x(x+1)$, we obtain

$$\frac{x+1}{x(x+1)} - \frac{x}{x(x+1)} > 0$$

or

$$\frac{1}{x(x+1)} > 0.$$

The numerator is positive. The denominator is undefined at $x = 0, -1$. On the number line we mark these critical values.

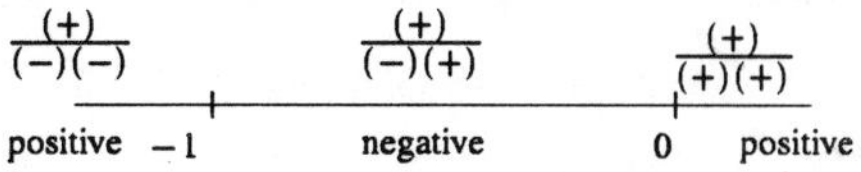

$$\frac{(+)}{(-)(-)} \qquad \frac{(+)}{(-)(+)} \qquad \frac{(+)}{(+)(+)}$$

positive -1 negative 0 positive

Therefore $\dfrac{1}{x} > \dfrac{1}{x+1}$ when $x < -1$ or $x > 0$.